COMMERCIALIZING INNOVATION

TURNING TECHNOLOGY BREAKTHROUGHS INTO PRODUCTS

Jerry Schaufeld

Apress®

Commercializing Innovation: Turning Technology Breakthroughs into Products

ISBN-13 (pbk): 978-1-4302-6352-4

ISBN-13 (electronic): 978-1-4302-6353-1

Managing Director: Welmoed Spahr
Acquisitions Editor: Robert Hutchinson
Developmental Editor: Matthew Moodie

Coordinating Editor: Rita Fernando
Copy Editor: Kezia Endsley
Compositor: SPi Global
Indexer: SPi Global

Distributed to the book trade worldwide by Springer Science+Business Media New York, 233 Spring Street, 6th Floor, New York, NY 10013. Phone 1-800-SPRINGER, fax (201) 348-4505, e-mail orders-ny@springer-sbm.com, or visit www.springeronline.com. Apress Media, LLC is a California LLC and the sole member (owner) is Springer Science + Business Media Finance Inc (SSBM Finance Inc). SSBM Finance Inc is a Delaware corporation.

For information on translations, please e-mail rights@apress.com, or visit www.apress.com.

Apress and friends of ED books may be purchased in bulk for academic, corporate, or promotional use. eBook versions and licenses are also available for most titles. For more information, reference our Special Bulk Sales–eBook Licensing web page at www.apress.com/bulk-sales.

Any source code or other supplementary materials referenced by the author in this text is available to readers at www.apress.com. For detailed information about how to locate your book's source code, go to www.apress.com/source-code/.

Apress Business: The Unbiased Source of Business Information

Apress business books provide essential information and practical advice, each written for practitioners by recognized experts. Busy managers and professionals in all areas of the business world—and at all levels of technical sophistication—look to our books for the actionable ideas and tools they need to solve problems, update and enhance their professional skills, make their work lives easier, and capitalize on opportunity.

Whatever the topic on the business spectrum—entrepreneurship, finance, sales, marketing, management, regulation, information technology, among others—Apress has been praised for providing the objective information and unbiased advice you need to excel in your daily work life. Our authors have no axes to grind; they understand they have one job only—to deliver up-to-date, accurate information simply, concisely, and with deep insight that addresses the real needs of our readers.

It is increasingly hard to find information—whether in the news media, on the Internet, and now all too often in books—that is even-handed and has your best interests at heart. We therefore hope that you enjoy this book, which has been carefully crafted to meet our standards of quality and unbiased coverage.

We are always interested in your feedback or ideas for new titles. Perhaps you'd even like to write a book yourself. Whatever the case, reach out to us at editorial@apress.com and an editor will respond swiftly. Incidentally, at the back of this book, you will find a list of useful related titles. Please visit us at www.apress.com to sign up for newsletters and discounts on future purchases.

The Apress Business Team

This book is dedicated to entrepreneurs who have succeeded and those who failed. Their experiences provide the joy of learning and the opportunity to improve.

Contents

About the Author

Jerry Schaufeld's wealth of experience in entrepreneurship, operations, and general management of technology-based companies ranges from his current role as Professor of Entrepreneurship and Technology Commericalization at Worcester Polytechnic Institute (WPI) to an assignment as a consultant at the technology transfer interface of Children's Hospital in Boston.

He served as Director of the RI Slater Fund, (a state-level investment resource for early stage companies), was President and CEO of Mass Ventures (a quasi-public technology incubator), and has a "hands-on" track records ranging from functional to board-level advisory roles in several early-stage companies. He is a founding member of the Launchpad Angel Group in Wellesley and co-founder of the Cherrystone Angels in Rhode Island. In addition, Mr. Schaufeld is an advisor to the Boynton Angel Group in Central Massachusetts, a charter member of the national Angel Capital Association (ACA), and a founder/participant in the regional NE Angels ACA group. Mr. Schaufeld was a founder and the first Chairman of the MIT Enterprise Forum, which now has 23 chapters around the world. He also founded the Incus Group, which is a CEO-level business acquisition and resource collaboration.

Mr. Schaufeld was co-founder and COO of Phoenix Controls Corporation. The company pioneered an approach to the control of air flow in critical laboratory and hospital environments. After ten operational years, it was successfully sold to a Fortune 500 company.

With a graduate engineering degree, research experience at MIT, an MBA, a professional engineer's license, and a Professional Board Director's Certificate, Mr. Schaufeld has a distinguished technical and operations-savvy managerial career. His current interest and research is in the area of improving the probability of success of early stage, innovative, technology-based ventures in their quest for commercializing innovative opportunities.

Acknowledgments

The complete list of important people is long. Some of the special folks who helped with the journey of creating this text include:

- Professor Ed Roberts at the MIT Sloan School of Management, whose unabashed enthusiasm for reporting on his research on MIT spin-off companies and entrepreneurship set the classroom stage that helped enable my career in early stage ventures.
- Art Parthe (deceased), co-founder of the MIT Enterprise Forum and Engineering Group Leader at Draper. His utter belief in "can do" set the pace for astonishing accomplishments that finally led to the development of flight-control systems capable of landing a man on the moon.
- Bob McCray, founder of Worcester Controls and former Director of my company, Phoenix Controls. Our many discussions about integrity and how the forces of growth are connected to the corporate lifecycle phenomena were absolutely invaluable.
- Rico Baldegger, colleague and Director of the School of Business (HEG) in Fribourg, Switzerland, who not only furthered my appreciation the unique Swiss perspective but also opened my view of the role of global competition.
- Mac Banks, former Director of the WPI School of Management, who 10 years ago took the risk of hiring me as an unknown entrepreneurship academic and provided incredible support for a new and exciting career.
- WPI's Professor Diran Apelian has provided an impressive and awesome example of what can be accomplished at the intersection of theory and practice in his stewardship of the WPI Materials Processing Institute. It has set an example of applied academic leadership for which my colleagues and I are grateful.

- The fourth grade teachers in the Houghton Elementary School in Sterling, Mass. Not only did they show me the importance of teaching excellence and integrity, but they also demonstrated the value of instilling learning skills in the early years. That certainly includes the basis for teaching entrepreneurship awareness in the early grades that's now in vogue.

Thank you to my friends, family, and colleagues, whose endless and polite inquiries about the progress of this book helped. Their interest and attention were certainly appreciated.

Special thanks to my wife Sue, whose continuous support (and cajoling energy) for this project was crucial. It would not have seen completion without it.

Thanks also to those special folks at Apress. Jeff Olson the former editor who believed in the topic and helped the publishing cycle start. Robert Hutchinson, who later carried the project to the finish line.

Very special thanks to Rita Fernando, of Apress, New York, whose saint-like patience and energy to my endless tactical questions is really the force that drove it all to a final iteration. Finally, thanks to Matthew Moodie who joined the team late in the game, but whose help is appreciated.

Introduction

We live in a time of unprecedented attention to the forces of innovation and entrepreneurship. Throughout the world, the combination of these two focused efforts affords us the opportunity to both compete and profit on a global basis. To enable this, the technology that allows us to benefit and the channels of information that can communicate these ideas to those who can utilize them, are both changing at warp speed.

Within this dramatic environment, the processes for improving the probability of commercial success of these ideas lags behind. This book offers a model for increasing the odds as well as the narrative that supports it. Much of the work presented is derived from a graduate course I teach in the Worcester Polytechnic Institute (WPI) School of Business entitled "Technology Commercialization."

Rapid change affords opportunities we could only have imagined a few years ago. The drama of the change in "technology commercialization" is awesome. New models of nimble and adaptive thinking anchored in solid business fundamentals are required to realize its benefits.

CHAPTER

1

Technology Commercialization

The Legacy

We live in a time of incredible technological change. It is characterized by significant dislocations of people, political infrastructures, capital formation, and material resources. It is fueled by significant investments in innovation and entrepreneurship at levels of public and private funding. The dimensions are global and empowered by enabling forces such as the Internet, Facebook, Twitter, Google, and a seemingly endless list of social media-based design and communication tools. They use a new vocabulary that enables rapid and international levels of collaboration. The implications of all this are just becoming understood.

The stage is set for this book's premise in the first chapter. It starts with a suspicion that we look to innovation and entrepreneurship as directions to confront declining manufacturing bases, shortages of raw materials, environmental concerns, and a significantly changing global mix of wealth distribution. The attention to innovation has produced many new ideas and visions for directions ahead. The attention and the methods for converting those ideas to commercial reality have lagged behind.

This book argues that commercializing technology can be accomplished more efficiently and with greater probability of success than it has been done before. Enabling tools and information are more prevalent than ever. Methods that worked in the past to bring technology to market in various forms no longer work as well as they once did. Even still, we persist in employing them.

What works better now, for example, is to seek ideas for commercialization from multiple sources. Then, to consider the alternatives for creating commercial pathways to get products and services to the market based on unifying goals and visions. To accomplish this we need to look at the overall process of moving ideas to market reality.

First, how did we get to this point? It can be argued that things evolved as they did based on a somewhat stable model of economic order. Industrial strength was centered in the United States and Europe. Material resources were consumed by these nations from what seemed a limitless supply. Much of the new wealth created was built on the backs of impoverished nations like China, India, and South Africa. Although the riches of this approach were unevenly distributed, the model seemed to work. Today, forces like unemployment, poverty, large gaps in wealth, insufficient healthcare delivery and education, and diminishing resources like energy (and its impact on the environment) also share the stage of world economic concerns.

We simply have to do better! Uneven distribution leads to an unequal capacity to change. Investments in education, Internet-based communications, and formal research and development are so disparate that they make it difficult to create new solutions.

Note Innovation and entrepreneurship are not sufficient to ensure future change. Business decisions must focus attention on the processes for bringing those ideas to commercial reality so that they yield the highest probability of success.

Books, articles, professors, and pundits worldwide expound constantly on the need to promote innovation and entrepreneurship. These, we are told, will help sort out a new world order and provide more wealth and opportunity for all. Perhaps, but in my view, the current level of innovative ideas coupled with less disruptive entrepreneurial initiatives may not be sufficient to provide a better economy for all. Incremental gains fall short when we need more dramatic returns on innovation. Educators, business leaders, and innovative change agents must focus on improving the probability of success of their projects so that usable ideas come to the marketplace.

The attention to innovation has brought with it a flood of well-intentioned resources into focus. It has also brought with it a commoditization of our attention. An example is a project I worked on while serving as a consultant to the Technology Transfer Office of Children's Hospital in Boston (CHB). It was observed that bringing ideas to the pediatric market would be enhanced by a collaboration among six leading pediatric hospitals in the country. We coined a name for this project and called it the Institute for Pediatric Innovation (IPI). To ensure that we weren't infringing on someone else's trademark, we asked our attorney if we could use that name. She researched the name and said it was okay, but she also reported that there were 96 organizations in the greater Boston area alone that used the term "innovation" in their titles—wow!

We have more innovative ideas every day but, paradoxically, fewer life-changing breakthroughs. We are hamstrung by business models that reflect a prior age and rely on quarterly performance rather than metrics of innovation and change. An example might be in the automotive industry where yearly model changes are characterized by changes in color, name, and incremental feature modification. Disruptive changes do not appear to be part of the model.

Many innovators believe the best way (and maybe the only way) to commercialize an innovation is to start a new company. As this book will show, there are multiple pathways to bring ideas to market. Paying attention to how ideas reach commercial reality, the focus of this book, can change the game for innovators, entrepreneurs, and customers. You will learn more ways to bring products and services to market, helping speed the process of innovation and providing the potential for the dramatic breakthroughs we need to keep technology and society both progressing. Best of all, you'll learn new models that promise improved probabilities of bringing your ideas to the market successfully.

Waypoints in Innovative Commercialization

As with all complex stories, the history of technology commercialization is not smooth and continuous. Innovative ideas have been with us since the beginning of time. Cavemen struck flint stones against rock to create fire in their caves. The fire was used for warmth and for cooking food. If fire were the only outcome, the story would have been lost in the noise of history. That there was not a complementary sustainable value proposition in food and shelter, the importance of flint and rocks would have been trivial.

In this story, certain historical "waypoints" are worth noting.

They will help us look at several models of innovation to see if there are "lessons learned" that will help us develop models for going forward. A starting point is the work of W. Edwards Deming, who has a method for improving the quality of our manufactured products that seems both innovative and disruptive. His work was generally rebuffed in America but later embraced by Japan.

W. Edwards Deming and Innovative Change

Let's start with W. Edwards Deming. Born in 1900, he was trained as a statistician. He practiced in both government and industry. His mantra was the application of a rules-based statistical approach to quality. While working with the Ford Motor Company in Dearborn, Michigan in the 1980s, he observed that Ford product components that were built in Japan performed better than those built in America. They also had fewer customer complaints.

While advising management at the Nashua Corporation's disk drive manufacturing plant in Nashua, New Hampshire, he learned that even hard-to-measure aspects like "soft" (hard to quantify) defects in memory disk production could be remedied through statistical quality control techniques.

Deming had additional successes on U.S. soil. But as powerful as his insights were, American Industry did not embrace his methods. Japan did! He applied his work to basic Japanese industries, such as the automotive and heavy machine tool segments. The "waypoint" in this story is stark: Deming's success abroad highlighted the apparent loss of focus by America's industrial giants, resulting in its loss of markets, manufacturing capacity, and innovation.

The United States had the largest capacity in the world to manufacture automobiles. Toyota was an unheard of brand forty years ago. It was perceived to produce inferior products. By embracing the innovative changes championed by Deming, Toyota became the world leader of quality-based cars and achieved dramatic market share penetration in a few years.

With increased quality in Japanese products came customer acceptance, and soon Japan became the hallmark of high-quality hard goods such as automobiles, appliances, and machine tools. Today, many Japanese manufacturing plants extol the Deming philosophy and tout statistical control charts, herringbone (supply chain quality maps), and direct shop-floor control over processes in their passion for quality. America and Europe are now catching up, but they have a way to go.

Toyota emerged as a leader in this race and eventually produced the Lexus automobile which, for the last seven years, has achieved number one status in the J.D. Power customer satisfaction rankings. During this period, Mercedes, which had been in the top position historically, slipped to number seven. American-built vehicles shifted to the middle positions. Is this as simple as Deming's influence? Probably not, but his influence and his compelling energy to create quality is seen as the enabling power to erode what seemed an unstoppable technological and industrial dominance by America and Europe. This discussion extends beyond automobiles to hard industrial goods to televisions (GE and RCA were replaced by Sony, Sharp, Hitachi, Toshiba, Mitsubishi, Panasonic, and Samsung). The formerly entrenched industry leaders have disappeared entirely from the horizon.

Why look back to the 1900s and the Deming era? It is simple. Doing so provides an extraordinary lesson about the importance of not only embracing innovative change, but also the value of bringing the benefits of those changes and the resultant benefits to their customers. A recurring theme is that innovation is not only a creative aspect of creative skills and change, but it also brings value perceived by customers that can be converted to financial profits and market share metrics.

Walmart Expands While Manufacturing Nothing: A New Innovation Model Emerges

So far it would seem as if the journey were characterized by technological or industrial innovative strengths alone. Hardly, as the next "waypoint" shows. Sam Walton started Walmart in 1962. Its corporate headquarters are located in Bentonville, Arkansas. Walton worked in the retail business (JC Penney) and was determined to provide the consumer with a better deal. His was a crusade to change the retail shopping model. Clayton Christensen at the Harvard Business School would have coined this a disruptive change, yet, the gains were all in the supply chain and going to market customer issues. Central to his model was the creation of "super stores" that would allow large movement of goods globally and thus give the company better purchasing power. Not only was the idea innovative, but Walton brought the model to commercial realization at a level that became a new standard of retail efficiency. Innovative change is not constrained to those driven by technology. Walton clearly shows it can happen in commercial "go to market" and with novel distribution models.

This approach has not been without controversy. Communities resisted the creation of "big box" stores because of the perception that higher wage local jobs were undermined by the lower wages that a Walmart offered. They also raised concern that local businesses located near Walmart stores would go out of business. But the company prevailed and continues to flourish.

Today Walmart has operating revenues of $470 billion and a net income of $17 billion. It employs 2.2 million people. Embedded in the Walmart model is the innovative use of Radio Identification Tags (RFID), which allow it to "track" customer sales at the moment of the transaction. If a consumer purchases an item off a shelf in New Haven, Connecticut, the computerized system notifies a manufacturing plant in Xianju, China of the sale. Multiple sales signal an increase in demand and the plant increases output. Note China's role in this. In addition to being an example of disruptive change, Walton reflects the accelerating loss of America's manufacturing dominance and spotlights our need for additional models of commercialization to compete.

Why focus on Walton in a text on technology commercialization? Walton was an icon of innovation. In many ways he personified the soul of what we hope to accomplish in our innovative endeavor. Bentonville is hardly Silicon Valley or Route 128. Walton drove his ideas to commercial reality. Far from a perfect model, Walmart did alter the landscape of how we do retail business.

Thomas L. Friedman, renowned author and writer for *The New York Times* focuses on Walmart in his award-winning book, *The World Is Flat*. He observes that Walmart doesn't make anything. It simply distributes and sells products made from all over the world and offers them at low prices. However, the lower wages in China became a significant lure for Walmart to sustain its economic model. In 2007 Annual Report, the World Trade Organization (WTO) reported that Walmart imported $27 billion dollars of goods from China and was responsible for 11% of the United States-China trade deficit. More important, it was responsible for the loss of 200,000 American jobs during this period.

Lexus vs. Ford

Certainly, there are multiple examples of the impact of innovation to measurable commercialization. An interesting one again occurs in the automotive industry, and it reveals another important aspect of economic dislocation. I use this example in my classes in Technology Commercialization at Worcester Polytechnic Institute (WPI). It immediately engenders a discussion on the economics of innovative commercialization.

Let's look at a mundane example: it is the assembly of the exterior door handles in the high-end Lexus Model 460 and the Ford Explorer. The former sells for over $80,000 and the latter for $25,000. The exterior door handles in both models are secured by two screws. According to the Bureaus of Labor Statistics 2012 labor compensation report, the labor rate for hourly workers in the Lexus Company's modern Tahra Plant in Aichi, Japan, is approximately $18 per hour. At Ford, the worker in its mature Chicago, Illinois, facility receives a "loaded" $75 dollars per hour (hourly wage plus benefits). What is so fascinating about this is that the Lexus products have clearly distanced themselves from Ford and other manufacturers in terms of perceived quality. The difference is partly explained by the impact of accumulated union negotiations and workplace regulatory considerations contrasted to a lean manufacturing environment. The real impact on the new global competition is significant. We must find a better way. The wage differential in various global sites may completely dominate the discussion of how ideas come to market.

This litany of "waypoints" suggests that the journey to the new global competition is a continual one comprised of multiple and incremental steps and changes. It is certainly not as simple as one day realizing that the players have all shifted on the international stage. It is a gradual and continuous process.

Enter Innovation

Focusing on the automotive industry for a moment allows us to observe one more important element in the journey from ideas to commercial reality. It is the role of innovation. America has the largest capacity to manufacture automobiles in the world. Yet, the innovative Prius hybrid-powered automobile was developed and manufactured in Japan by Toyota. America simply didn't "get it."

The Prius is possibly the first significant innovative change in the automotive industry since the Otto cycle engine was shown to be an adequate source of propulsion. The early engines were introduced in the late 1800s. Fuel consumption was not even considered a design criteria. In the 2000s this changed dramatically. The Middle East exerted its ability to regulate oil production and flow to the consumer. That, naturally, boosted the price per gallon/liter considerably. As these economic tides shifted and the consumer's attention to fuel efficiency (and the environment) occurred, Toyota launched the Prius. It features a gas/electric propulsion system that allows the driving load to be shared by two different technology systems—the internal combustion engine and the electric motors. Forty miles per gallon became the norm. Today there are over one million Prius vehicles on the roads in America. Europe pursued a different path toward efficiency with diesel engines, but these are not as novel and innovative as the Prius. Innovation, in this case, was aptly rewarded with a new and robust market share.

Where was Detroit in all of this? It seems they were bound to the status quo. Perusing the annual reports of the "Big Three" manufacturers in the early 1960s reveals modest R&D expenditures (as a percentage of sales) amounting to 2–3%. By contrast, Silicon Valley enterprises are in the 9–11% range.

If innovation can be simply characterized by change of technology, design, speed, cost, and social impact, perhaps the greatest evidence can be seen in the applications of computer technologies. The list of examples is nearly endless, but are best seen in several broad categories.

The first is in the somewhat invisible use of computers in what is referred to as "embedded" applications. This includes the automobile. The average number of processors in a modern vehicle equals about 17. Processors are used to control engine performance, environmental concerns, radios, lights, and even seat locations. Other hidden (embedded) applications include appliances, thermostats, medical devices, cameras, musical instruments, and many others. In most cases, the performance of the application improves and costs are reduced—wow. In addition, these changes happen so quickly that they outstrip the normal product lifecycles of the technology.

A second category of change engineered in computer applications is through use of computation in the hardware and software that extends the use of the devices. Included are speech recognition, graphics- and sound-based applications (music), and games. Of course, the use of computers in information technology and data processing is enormous and expanding. Within the space of computer science, there has emerged a discipline called *data mining*. It is derived from the analytics of large data business tools of the 1960s, such as the use of Bayesian thermos and regression analytics. As the amount of data we create and use increases, better techniques and computer-based tools that have increased computational capacity have emerged. Data mining allows the data to be analyzed to find patterns and applications.

An extension of this category is in hardware applications like robotics, automation, and rapid prototyping. Each of these requires significant computational capacity that is faster, more technically competent, and driven by lower cost. Venture-backed commercialization projects were based toward computers and their enabling technologies. Investments were in semiconductors, disk drives, displays, and minicomputer (that is, Digital Equipment Corporation—DEC) manufacturers. Today there has been a significant change to software and applications firms.

The last—and somewhat overwhelming—use of computers is social media. Facebook, Google, Twitter, and the use of MP3s (for music) are some examples. Simultaneously, the emergence of the Cloud as a real-time storage and program source is accelerating. Pundits of the industries they serve agree that all these examples of innovation represent just the beginning of the new wave of applications and aren't the long-term game changers. New social, political, and legal applications are reported each day. Apps are bountiful and seem to overwhelm the existing capacity to fund and staff the rate at which their organizational models appear. It's truly the Wild West of technology applications. The ease of entry for these new applications is both good news and not so good news. On the plus side, almost every conceivable application can find market segments to serve. On the other side, the ability to create sustainable value propositions seems elusive. Traditional barriers to entry are nonexistent. Either the secure intellectual property assignments or the long-term sustainable value advantages seem elusive and thus distant to the traditional technology-based investments.

Innovation Practitioners

Although we may revel in the accomplishments of the individuals who successfully brought new ideas to commercial reality, it is also important to realize the role of the infrastructure and institutional models that support the process of commercialization.

As innovative solutions emerge as a national and priority, there are many examples of how the U.S. government is supporting change. One example is the Small Business Innovation Research (SBIR) program. Its goals are:

- To spur technological innovation in the small business sector
- To meet the research and development needs of the federal government
- To commercialize federally funded investments

Roland Tibbetts, the program founder, was quoted in an appeal to Congress for continuance of the SBIR (May 28, 2008) as saying that "ideas, however promising, are still too high risk for private investors, including venture capitalists." A hallmark of the program is that it disburses increasing amounts of money to promising projects in three phases. Funds increase as the idea progresses from idea to commercialization in measured phases. Annual disbursements exceed $100 million and are distributed through 11 federal agencies. Impressive technology-based companies, like Qualcomm, da Vinci Surgical Robots, and iRobot (Roomba robotic vacuum cleaners) were started with this source of funding.

One ponders the role of government in this process. Traditional risk models guiding venture and angel investors seem unable to reach into the early-stage potential. Although there are multiple successful examples, is it the role of public funding to assume that risk? The dimensions of current funding are enormous. For example, in addition to the SBIR funding models, there are also other, similar programs in the United States. such as the Small Business Technology Transfer (STTR) program. That program also disburses $100 million in funding for specific technology-based projects.

The corporate world has its share of contributors to innovation research. Certain companies have invoked a culture that allows them to sustain the model of innovation. 3M Corporation in St. Paul, Minnesota is an example. For more than 100 years, it has continually introduced new and disruptive products and services to the market. Wet and Dry sandpaper, Scotch Tape, and Post-it Notes are just three examples. They were not perfect in how they came into fruition. Post-its, for example, came from a manufacturing defect of a batch of substandard adhesive that was about to be discarded. It didn't seem to stick very well. The product was almost stopped internally until an administrative assistant used it for her own short-term note applications that utilized the short adhesive properties and validated the product's to management. Today, it is a standalone industry. At least one element of the commercial cycle allows for intuitive and opportunistic events. It is not in contradiction to more disciplined approaches, just an allowance for it on an array of potential project offerings.

Apple's Innovative Culture

Although 3M examples have been a consistent benchmark in the story of innovation, today there is a data point that overshadows most others. In a world that is characterized by accelerating and rapid change and innovation, the personal computer and complementary electronics continue to lead. Within that group, Apple stands out. Started by Steve Wozniak and Steve Jobs, its original vision was to deploy innovative, user-friendly computers to the average user. The two constantly challenged themselves and others to be innovative in their design and anticipate customer demand. The goal was to offer solutions that exceeded consumer expectations.

The Apple II was an instant hit. While its popularity could have been limited to educators and hobbyists, it benefited from the unrelated development of VisiCalc, a precursor to Microsoft Excel. Dan Bricklin, the inventor of the software, observed a professor at the Harvard Business School struggling with rows and columns of information in accounting problems and developed the VisiCalc spreadsheet software for the Apple II. It enabled the machine to have a useful purpose in accounting and numerical analysis.

Wozniak had designed the computer to use a mouse and revolving text line that could wrap around the edge of the screen. It was originally developed by SRI. It even had two floppy disks on board for program and data access. Great attention was paid to the outside design to allow the user to feel the natural aspect of the man-machine interface. By modern standards, it was a primitive machine—but that may be one of the fundamental characteristics of innovation. It is that the effort to bring an idea to the marketplace of the users is the heart of the innovative process. Once an idea sees the light of day, incremental and disruptive changes enhance the idea and allow it to find greater utility. Maybe the analogy of the oyster that constantly polishes a grain of sand to produce a perfectly round pearl serves an illustrative purpose here.

Walter Isaacson, in his 2013 biography of Steve Jobs (same title), points out that the next steps for Apple and Jobs were not as smooth as a good story would predict. Jobs has a fallout with the co-inventor, Steve Wozniak. The follow-on model of the Apple II, named Lisa, didn't achieve market acceptance and the company faced bankruptcy. Three outside CEOs were brought in and soon dismissed. Finally, Jobs himself was fired. He went on to co-found a Silicon Valley animation company called Pixar and began work creating computer-based technology for the movie industry.

In a somewhat abrupt reversal, Jobs returned to Apple and implemented a series of dramatic and disruptive changes. Isaacson goes into great detail about the effect of Jobs's reentry into the company. From demanding attention to product detail to abusive interactions with employees and investors, Jobs emerges as a hero for introducing multiple major innovations. The company name changed from Apple Computer to Apple. Products included the iPod, iTunes, iPhone, and the iPad—each of which was to transform the industry segment to which they were committed. The iPod transformed the way we buy and listen to music, and the iPhone set the pace for handheld communication devices and the creation of apps. The iPad pioneered a new class of laptop computing. The results were significant. Apple enjoyed the largest market capitalization of any U.S. company and largest corporate cash reserve in history. Tragically, Jobs died of cancer in 2011 and a professional team now tries to continue his legacy.

Historians and academics will judge whether this company will continue to sustain its value after Jobs's untimely death. Early and current decline in the stock price may be a continuing trend. Competitors like Samsung have been able to dilute Apple's market share by penetrating both the iPad and iPhone segments. Whether it sustains these gains or not, it's quite clear that Apple set a new standard for corporate innovation.

The Bose Corporation

Amar Bose offers yet one more perspective. It lies in the ability of a closely held (private) company to invest beyond the competitive standards of its peer companies. Trained as an MIT Physics professor, Bose started his company, called the Bose Corporation, with $10,000 borrowed from his thesis adviser. His goal was to create high-performance products like superior acoustic speakers, noise-canceling headphones, and more. His company was private and substantially owned by him. This allowed him to pursue certain directions in product development that he didn't have to get approved by shareholders and undergo the scrutiny of other investors.

Bose passed away in July of 2013. *The Boston Globe* obituary indicated that he had invested over $100 million in a new acoustic automotive suspension system. Even though the Bose Corporation has reported sales in the multi-billion-dollar range, it is hard to conceive that much money could be applied to a "pet" project. The story reveals how a closely held company can reach to areas of research resource allocation not available to public institutions, which operate in the sphere of regulatory and shareholder oversight.

Curt Carlson: The SRI Approach

There are many other practitioners who personify the innovation process uniquely. Curt Carlson is an example. Curt pioneered the application of HDTV and other optical image quality innovations. As head of the SRI International think tank, he led an organizational model of innovation. In his book *Innovation*, he delineates a process called NABC to help individuals coordinate and maintain discipline in the innovation process. NABC stands for:

- N—Needs (as defined by the customer)
- A—Approach (to market)
- B—Benefits (per cost)
- C—Competition

At Worcester Polytechnic Institution (WPI) in Worcester, Massachusetts, where Carlson is a trustee, he helped introduce the NABC process as part of the commercial justification of the student-based research projects they have to complete as part of their undergraduate curriculum.

Innovative Thinking: Clayton Christensen

Clayton Christensen, a Harvard Business School professor and author of many books on the nature of disruptive innovation, also observed that disruptive change happens at a certain rate. In his earlier work, he focused on the disk drive industry, where a rapid change of technology had a major impact on the 35 or so manufacturing companies in the business of creating product. The changes dealt primarily with the size and recording capacity of the disks used in the industry.

Early in the game, IBM pioneered a computer memory platter that was 14 inches in diameter but had capacity for only a mere 5 megabytes of memory. Later drives went from 14 inches to 8 to 5.25 and finally to 3.5 inches diameter in a space of a few years. Each generation increased the capacity of the disks by algorithmic amounts. Today, at 1.6 inches in diameter, the capacity is over 1 terabyte of information. Drives are so small that they now fit into cell phones. Just as important, the product development cycle shrunk from 2.6 years to less than 6 months.

That rate of change had enormous impact on both the producers and users of the drives. They simply couldn't transition from one product range to the next fast enough. Of the original 35 companies, there are now only four. Today, companies are judged not only by their financial metrics, but also by softer ones like how well they can adapt to shorter lifecycles and market changes.

IDEO: Innovation in Design and Consulting

The roster of astute innovation practitioners reaches beyond icons like Carlson and Christensen. Organizations of consultants such as IDEO serve as an example. IDEO was formed by David Kelly and others out of Stanford University in 1978. One aspect of its mission is to "help organizations build the capabilities required to sustain innovation." Embedded in their methodology is a process called "design thinking" in which problems are envisioned as overlapping spaces identified as inspiration, ideation, and implementation. This unique Venn diagram approach stands in contrast to a serial, inline, and step-by-step approach utilized by many problem solvers. IDEO also uses models called "road maps" to demonstrate the future implications of decisions and concepts being considered by their clients.

Innovation at Foundations

Organizational models also reach into the world of foundations. Classic among those is the Kauffman Foundation located in Kansas City, Missouri. It was created by Ewing Marion Kauffman in the mid-1960s. Mr. Kauffman was the founder of a world-class organization called Marion Labs. The foundation has assets valued at over $2 billion and is one of the largest private foundations in the world. The focus of its work is in entrepreneurship and underserved populations such as women and minorities. They also promote innovation. Their goal is to foster a "society of economically independent individuals who are engaged citizens that contribute to the improvement of their communities." Included are programs for innovators in science and technology.

Certainly, the path to foundation involvement in commercialization doesn't stop at large, private organizations. An example is the Kern Family Foundation of Waukesha, Wisconsin. Robert Kern, the founder of the foundation along with his wife Patricia, started a company called Generac Power Systems in 1959. The company designs and manufactures backup power supplies, primarily engine-powered generators fueled by natural gas. In 1998, Mr. Kern sold a division of the company and created the charitable foundation with the proceeds. The mission of the foundation was to instill the "entrepreneurial (innovative) mindset" in young engineers. Embedded in this work is an innovation program entitled the Kern Entrepreneurship Education Network (KEEN). It is comprised of over 20 universities and is characterized by multiple seminars and workshops in the area of hands-on innovation. At WPI, for example, I created a very successful course entitled "Innovation and Entrepreneurship for Engineers" with such funding. It is presented in multiple offerings per year. Each is oversubscribed to the level that students are routinely turned away from registering for the course.

The Pattern Emerges

A pattern emerges from this array of "waypoints" to innovation. It is that there are many paths and they lead to multiple sources of inspiration. Without a focus or a defined goal, the design and development efforts, as creative as they might be, fall short of presenting new opportunities. This was summed up by Buzz Aldrin, the Apollo 11 crew member who walked on the moon, in a cover article of *MIT Technology Review*. He lamented, "You promised me Mars colonies. Instead, I got Facebook."

In the same article, Max Levchin, a co-founder of PayPal (the Internet-based credit card payment company) observes that "I feel we should be aiming higher. There is an awful lot of effort being expended that is just never going to result in meaningful disruptive technologies."

Some point to the venture investment community as the basis for the shift from bold innovative projects to incremental ones. According to Bruce Gibney, partner in a Silicon Valley venture firm called Founders Fund in an AOL TechCrunch interview with Alexia Tsotsis (December 2012), "In the late 1990s, venture portfolios shifted away from funding transformational companies to companies that solve incremental or even fake problems. Venture capitalists have shifted from funders of the future to become funders of features, widgets, and irrelevances." Twitter gives 500 people jobs for the next decade, but what value does it create for the entire economy?

Entrepreneurs like Bill Gates, Microsoft's founder, were determined to "put a computer on every desk and in every home." Steve Jobs of Apple wanted to make the "best computer in the world." Whether there is a shift in investment strategies, or entrepreneurs choose incremental problems, there is agreement that we no longer tackle challenges of the magnitude and impact of the Moon or Mars projects.

It is not for a lack of agreed-upon issues that we seem to have lost sight of meaningful projects. In the National Academy of Engineering's (NAE) 2012 Annual Report, seven "Great Problems" were identified. A billion people still want and need electricity, millions are without clean water, the climate is changing, manufacturing is insufficient to create enough meaningful jobs, population growth outstrips our ability to deliver food and nutrition, education is becoming a luxury, and diseases like cancer or dementia will strike almost all of us.

These problem are as interesting and challenging as the Mars mission, but somehow we no longer find focus in their solutions. Whether our lack of progress is the result of investment strategies, political actions, lack of entrepreneurship, or educational problems, the situation is ripe for change.

In the rest of this book, I will outline the steps that can be used to regain our footing and solve the big problems through the process of commercialization and how it can adapt to the needs of a new world of global competition, changing (and declining) resources, and dramatically changing technology.

I believe a pattern has emerged that suggests there are sufficient innovative ideas and solutions, but they lack the directions and process models to enable them to help the world. This further suggests that process innovation may carry as much weight as the actual solutions themselves. When there was a seemingly endless supply of resources and capital, the attention to process may have been secondary; now it may be the central path for moving forward.

Note Incremental change consumes significant resources of capital, time, materials, and human effort. Attention to bold, disruptive issues such as solving problems of energy, food, healthcare, and sustainability can be as challenging and rewarding as the trip to the Moon or even Mars.

Beyond process lies the challenge of what I call the areas of "Big I" and "Little I." Big I encompasses a range of innovations that are bold, disruptive, and change the course of the world in which we live. Putting a man on the moon was certainly one of them. 200,000 apps used by Apple aficionados might be at the other end of the innovation spectrum.

Little I is all about incremental changes, improvements, and applications that have a marginal effect on our lives. Although they serve as product and market extenders, they consume significant human capital and financial resources to accomplish. Clearly a better use of our resources is toward substantive projects. Defining a strong and enabling value proposition helps focus on the best allocation of resources for these tasks.

Summary

The context in which all of this transpires is set against a backdrop of change that is so rapid it can't be ignored. Increased population, diminishing resources, and breathtaking connectivity driven by computers and social media argue that we must find a better way. And we must do it all better. The model described in the following chapters offers a path to bringing practical solutions to commercial reality faster and better and with a higher probability of success. As you'll see, it brings the reader back to some basic business fundamentals in its implementation.

CHAPTER

2

The Commercialization Model

The Value of a Process

In the first chapter, I suggested that there is an abundance of ideas, projects, and even companies to fulfill our need for innovative and entrepreneurial alternatives. Yet, the economic measurements of employment, financial growth, and wealth distribution lag behind this and suggest that we need to find even more sources. I looked at the disposition of the creative energy and output needed to achieve success in these efforts and realized that there may be a better way to focus on the commercial opportunity those ideas have. It is to learn how to turn those ideas into commercial activity. A process (model) was called for!

A model allows for the effective means of sharing ideas, and it serves as a platform for measuring progress and outcome effectiveness. It is based on the idea that the probability of success of the outcome can be improved by process discipline. It results in a more efficient allocation of resources and a more efficient and timely access to user markets.

Using a model or process methodology offers a more subtle advantage. Because the elements of the traditional process of commercialization tend to lose the nimbleness and flexibility needed for improved outcomes, a more vibrant approach is needed. Utilizing a visual, agreed upon central model constantly reminds us that there are alternatives at each step.

Let's look at an example. It involves choice of pathway to exploit new ideas. Typically, startup models are used to begin the process.

Nimble behavior includes the lure of startups as a means for exploiting commercial success. Certainly, there are advantages of a startup alternative. They include a clean start, focused objectives, and the upside potential of equity participation. The excitement of a successful new enterprise is the ultimate draw. Yet, in many ways, it may be the most unlikely path for commercial success. There are alternatives to reaching the market in ways other than through startup ventures. The process model I present in this chapter offers a disciplined, more nimble approach to selecting the most probable effective pathway to commercial reality. It may not be a startup!

So Why a Model?

Certainly, the following is not meant to be treatise on business modeling. This chapter is, however, a platform for looking at business models as tools for delineating and understanding commercialization processes. Even more relevant is to see if the business model methodology can improve the probability of success of the outcome.

Jarkko Tapani Pellikka and Pasi Malinen, in the December 16, 2013 issue of the *International Journal of Innovation Technology,* published an article in which they argued that particularly in small technology firms that face increasingly severe competition, only effective (and documented) commercialization processes can secure survival of the venture. Additionally, they observed that smaller technology-based firms seemed to pursue more closed approaches to development, leaning heavily, for example, on R&D. In these firms there is little margin for error nor deep resources to start again.

What follows is not an absolute solution, nor is it a "one size fits all" static tool for improved success. That would be too simple and not reflective of the dynamic and innovative nature of commercialization. So, let's look at what it the model is and what it does:

- In its simplest form, it is a graphic representation of multiple inputs to a process. It can represent collaboration and consensus in its formation and implementation. It also invites change as the information it requires to form decisions evolves.
- Effective models have the component of "learning" that improves outcomes through continuous feedback.
- It helps break down complex process elements of commercialization so they can be distributed to others and understood in their own right. The elements

of the proposed model shown in this book will be first identified and later expanded in subsequent chapters.

- It invites measurement or quantification, so that the effort can be benchmarked to other projects and industry standards and thus be maintained. Once measured, the effort becomes an opportunity for continuous improvement by setting forward-looking metrics to be invoked. From this comes the final calculation of how the elements of a given project contribute to the probability of success, which is the end goal of positive commercialization processes!
- A model invites a discipline similar to a pilot's preflight checklist. Even though pilots fly similar planes and similar routes, they always revert to an orderly checklist before starting or committing to the flight resources that the journey requires. It still amazes me how many projects are cited that have not been subjected to the rigor of a discipline similar to the pilot's process or its equivalent.

Despite its many advantages, the model development is not without faults. A partial list might include:

- By adhering rigidly to a model's constraints, the model can be self-defeating in that it produces actions using expertise in the model's workings and thus encourages users to lose sight of its vitality. Using the pilot's nomenclature again, we can allude to a term called situational awareness whereby the pilot makes decisions in the context of all that is happening to the aircraft at any one moment. Later we will look at workarounds that might overcome this issue.
- Although the predominant theme of this section is to argue that a strict, disciplined model serves an organization best, there needs to be a balance in how it is implemented. Certain computer tools, such as dashboards and other visual presentations, can present a pictorial representation of the variations of ideas that can then be compared to strict rules. It could also be argued that some small range of projects would be allowed rules other than those that fit the unifying goals.
- The model might be constructed in a manner that simply does not reflect the corporate (or project) goals or vision. Similarly, it may lack the ability to reflect the organization's inherent capacity. Either is a recipe for suboptimal results.

- The model simply yields poor outcomes in a variety of metrics, including profits, market share, time to market, or other perceived measures of growth.

With a context for business models in hand, it becomes relevant to look at those issues that surround the process of commercialization. The basis for this is derived from graduate course that I teach at WPI in Technology Commercialization and discussions with product development and new venture professionals.

A Model for Commercialization

An early observation was that discussions about process were focused on specific issues such as the mechanics of funding or feasibility analysis.Very little was mentioned about the overall process. As mentioned later in the book, I consulted with a team at the Chrysler-Fiat research labs that focused on new product development. Almost consistently they focused on the issues of Opportunity Recognition and how to overcome the current (and later identified as chronic) logjam of more than anticipated flow of unsolicited proposals. If only that backlog could be overcome they felt they could improve overall performance.

In reality, failure of any element in the process can be a chance for the overall initiative to go astray. There are at least five separate sections to the generalized steps of the cycle. Cleary the number of steps will vary with the sectors, stages of growth, or the amount of capital required. For example, the regulatory impact of decisions in biotech or large capital equipment projects like a steel mill enlarge the model beyond a simple startup. The question now becomes how to define a generalized model that can be used as a foundation for more elaborate applications.

With these caveats in mind, let's look at a possible graphic model for portraying a cycle of commercialization (Figure 2-1).

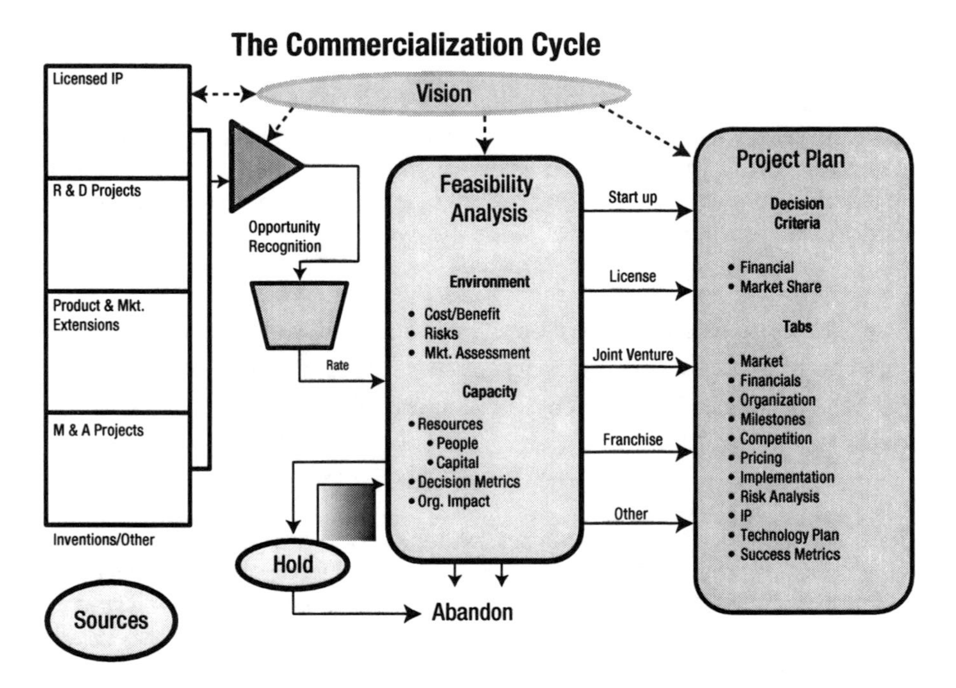

Figure 2-1. The commercialization cycle

While not meant to be inclusive, this somewhat elaborate presentation outlines a path of decisions and processes that describe the cycle of how ideas move from concepts to the reality of commercial viability. It suggests that if the disciplines outlined were followed, it would yield a higher probability of success as measured by defined metrics of market financial driven measures. The model is comprised of the following sections:

- Sources of ideas
- Opportunity recognition
- Feasibility analysis
- Going-to-market methods
- Project plan

What follows is a brief introduction to these sections. Subsequent chapters explore these topics in detail.

Sources of Ideas

Finding ideas worthy of becoming projects for successful commercialization is a daunting yet critical part of the entrepreneurial journey. As suggested in the last chapter, it isn't that there are not enough ideas. Rather, it is an abundance of possible avenues for achieving the project's goals that may confuse the selection process. Indeed, this is where the cycle begins.

Note If a commercialization project arises within an existing organization, it becomes important to align a project proposal with the corporate vision and its stated goals. When you can couple the project directly to the stated vision, it will make it much easier to secure financial, human, and other resources required from the parent.

When sorting out multiple sources of innovation, it's important to avoid blindly repeating the use of singular source channels. Actively and dynamically seeking out the best match of source to a given project becomes critical. In the graphic shown in Figure 2-1, you see five traditional groups of resources for technology commercialization. This is not meant to be an inclusive list, but rather a partial selection of alternatives. It includes "other" as a placeholder for newer and unconventional sources that may apply to a given project.

Aircraft pilots who are certified to fly airplanes in conditions that compromise outside visual references—such as when flying through cloud banks—follow Instrument Flight Rules (IFR). In their training for being certified to do, pilots learn to use six primary instruments to synthesize aircraft performance in such conditions. Training includes the skillset of scanning the instruments for multiple information inputs in real time. If the scan is broken and the pilot becomes fixated on any one instrument, it becomes a recipe for failure and a loss of control follows.

This parallels the quest for business opportunities quite well. Fixation on any one path invites failure.

It is tempting to look to invention first. Much has been written about creativity and role of developing fundamental ideas. Most of us can conjure an image of Edison diligently working over hundreds of lamp filaments in his West Orange, New Jersey and laboratory. One can relish the "a-ha" moment of invention.

Yet when we look at it on a model of probable outcomes, in-house invention is perhaps the least likely avenue to success. That's why Figure 2-2 starts with looking at licensed technology and its inherent strengths and then encourages similar scrutiny from the other sources. (Chapter 3 examines each of the following category sources in more detail.)

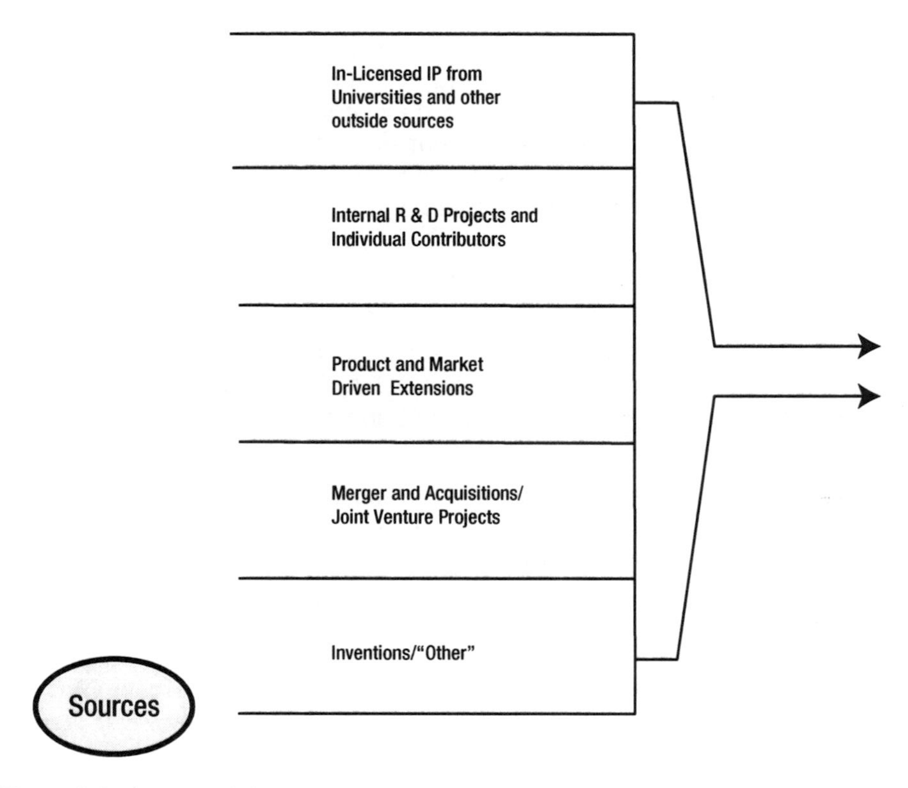

Figure 2-2. Sources of ideas

Opportunity Recognition

Much has been written about how ideas are sorted out or distilled in the context of resource-allocation decisions. Simply stated, academics (and others) refer to this process as ***opportunity recognition***. There is not much agreement in the literature about the various approaches to the process. Much of the decision processes default to analytical tools and approaches over intuitive approaches. Intuition has been significantly discounted. However, it is possible that intuition plays a larger role than we thought and needs to be looked at.

The Role of Intuition

Although a major premise of this text centers on the quest for systematic approaches to commercialization, it becomes important to acknowledge a "placeholder" for intuition or a "gut feel," as it is sometimes called. Individuals with professional disciplines such as science and engineering reach for a more orderly process.

As decisions move upward in corporate hierarchy, their nature becomes increasingly more ambiguous and generalized and less susceptible to detailed analysis. The CEO who must drive the elements of a corporate vision as a basis for decisions lives in a very different domain than the middle manager who might be deciding to purchase a major piece of capital equipment. The latter's metrics are more formula-driven and have succinct values and data to drive their decisions. Whether based on financial metrics like returns on investment or market penetration measurements, or even productivity gains, good decisions are driven by analysis.

Intuition and intuitive decisions are characterized by a range of scope and experience. Figure 2-1 attempts to show the myriad of factors that influence intuitive decisions as well as the range of expectations about the validity of the outcomes.

Another perspective about the value of intuitive decisions was amplified to me during a visit I made years ago to the Gibson Guitar division of Wurlitzer Music. It is located near Buffalo, NY. During a tour of the manufacturing process I noticed a somewhat senior individual playing each newly minted guitar and then signing his name to it. The process for manufacturing the instrument was quite elaborate. This step was at the end of the production process where the device under scrutiny had absorbed its maximum cost. Yet, this one individual became an important checkpoint and last (intuitive) step of manufacture. He was a quality control inspector. In spite of all the apparent automation and control, this one person relied on a well-honed intuitive feel for the quality of the instruments. It was his signature that appeared on the inner surfaces of each musical instrument.

A Disciplined Approach

I was once in a meeting with the Director of Innovation at Chrysler Corporation. He said they receive thousands of ideas and proposals from their employees, but that there was not an agreed-upon process within the organization to decide on which ones to move forward. They utilized a committee voting process but in reality, he agreed, it required a more methodical approach to improve.

What follows is an exposition of a process developed at WPI in a student major qualifying project (MQP) thesis. Among other things, it demonstrates a selection accuracy of one out of eleven. That means that when presented with 11 projects, the model can ascertain which project is most aligned to the organization's vision. This resolution is important as it sets the stage for only considering projects that are closest to the overall vision and goals of the project or the company. Literature suggests that one out of five was considered adequate resolution.

At the time of the thesis existing approaches seem to peak out at a resolution at the one to five level. That means the methodology can only determine of one out five projects considered meeting the project criteria. The implied accuracy of one out of eleven is a marker to the effectiveness of the WPI approach. A higher resolution suggests that the process is more accurate. To extend this idea the overall objective of this book is to see if the probability of success can be improved. Better discrimination at the early part of the process suggests a more probable positive outcome.

The student who led the project, TJ Lynch, was a senior who wanted to pursue an entrepreneurial career upon graduation. Like many ambitious and curious students, he felt unable to decide which course of action to follow. He commented that "all of them looked good." As his co-advisor I suggested that a more meaningful direction would be to develop the model in the context of selecting a project from multiple opportunities. He agreed and set the stage for the MQP and the research of project selection.

The process begins with an orderly examination of the goals of the project and its "vision." That vision statement is broken into elements that will be quantified and used as weighting factors (multipliers) in the assessment process. For example, if the overall objective is to be a low cost, technology leader of widgets, larger weights (multiplier percentages) will be allocated to those projects that present widget technology and cost-effectiveness. These multipliers will then be applied to the assessments of the individual functional areas.

The actual project is then broken into functional elements, and the weighting factors are multiplied and summed to provide a project ranking. The details of process are delineated in the WPI MPQ entitled, "The Development and Application of a Multiple Analysis Tool for Entrepreneurs." It is a public document and can be accessed through the WPI Library System.[1] A detailed look at the process is contained in the report and will be further examined in Chapter 4.

There is a need to separate an opportunity from an idea. An opportunity can be held to a fairly rigorous set of tests of viability, financial metrics, impact on organizations, and so on, while ideas seem to fall into large buckets of "good" or "bad." The Lynch model concentrates on defined opportunities and the need to quantify them in the context of their definable goals.

Again, there is a dichotomy between measurable attributes and a more subjective intuitive approach. In my many discussions with project managers and new market development mangers, they favor a consistent quest of more analytical approaches.

---

[1]https://www.wpi.edu/Pubs/E-project/Available/E-project-050509-091115/unrestricted/tjl-MQP-MOpA-Tool-D09.pdf.

Referencing the Andicivili, Cardozo, and Ray 2000 article published in the *Journal of Business Venturing*, the authors focus on a concept of "Entrepreneurial Alertness" as a driving force for creating opportunities. Mr. Robert Kern of Waukesha, WI was the founder of the Generac Company. At retirement, he sold the company and used part of the proceeds to create the Kern Family Foundation, which is driven to help engineering schools instill the "entrepreneurial mindset" into their engineering disciplines. I was a Principal Investigator (PI) recipient of a Kern Foundation grant designed to develop courses in innovation and entrepreneurship for engineers. It is constantly oversubscribed. If these directions might be considered a continuum, then certainly Steve Jobs' penchant for customer-driven design would be an end point.

Is It Real? The Role of Feasibility Analysis

Is the project (or product) real? Should we do it? Is the right time? Are we ready to do this? Are these the proper questions of uncertainty, or are they the proper assessments of whether to commit resources (both human and financial) to a given project?

The answer is, of course, "it depends." Many sound decisions are made on intuition and the inertia of "we've always done it that way!" In a time when competitive global pressures and rates of decreasing technology lifecycles prevail, maybe it's a signal for the pendulum of intuition versus method to shift toward orderly assessment processes.

Best practices in feasibility analysis fall into two broad categories. The first is the metrics of the opportunity as expressed in terms of market size, competitive forces, technology status, and consumer trends. The second category focuses on the readiness of the organization or project to commit and compete in the given space. Readiness is measured in balance sheet ratios of cash and liquidity, people in terms of availability and skillsets, and large capital metrics like available space and capital investments.

One of the great benefits of a disciplined feasibility analysis is the ability offered to the initiating organization to abandon a project or put it into a dynamic holding pattern that allows the proper conditions to be in place before proceeding.

The feasibility analysis issues will be further examined in Chapter 5.

Going to Market

Clearly, the domain of marketing and gathering the resources required for achieving market presence is the world of professionals. There is another dimension of going to market that is strategic and determines the path of commercialization. It is the set of decisions that determine which channel best serves the given project. Like the issues surrounding the sources of ideas where fixation on a singular channel dominate the thought process to its detriment, putting on your blinders when it comes to a go-to-market strategy is equally harmful.

Going to market via startup may be a good direction. Startups are most intriguing. Somehow the lure of a new venture complete with the excitement of the equity's upside potential is a most glamorous alternative. Yet somehow it is hard to imagine that the risks and fragile dynamics of a startup make it the uniformly best platform for exploiting market presence. As the model shows, there are robust alternatives in licensing, joint ventures, and franchises that you need to consider. Each has its own positive and negative characteristics.

Marketing is the rich portal of bi-directional information that flows out of the organization and back into it. Advertising, pricing, promotional efforts, and trade shows were the traditional domains of marketing. Social media and the Internet have accelerated these channels at a rate that is hardly understood. These classical functions had tragically become confused and intertwined with the sales function in many companies. It becomes management's responsibility to carefully delineate the two. Beyond these traditional aspects of outward information, marketing becomes the critical strategic lifeline for competitive information and is product feedback for later stage improvements. It also becomes a general bellwether for market and customer trends. A sharp CEO or senior manager learns to rely on the feedback and information. Options for going to market will be explored in Chapter 7.

The Project Plan

I suspect that as early as the building of the Egyptian pyramids, there have been project plans to delineate and chart the path of complex human endeavors of all sorts. In recent times, there is renewed interest in project planning for significant undertakings such as putting an astronaut on the moon or even on Mars. Constructing large buildings or building technologically advanced aircraft certainly means utilizing careful planning techniques.

What has changed over the centuries are the tools and the sophistication of project-planning concepts and their enabling platforms, such as computer-driven models. Broadly, these tools fall into two categories that embrace decision alternatives and performance metrics.

One of the earlier decision tools is the Program Evaluation Review Technique (PERT). It was issued created by the U.S. Navy to help manage the complex Polaris submarine project. It featured a series of nodes and vectors to represent the tasks and the time required to finish them. A graphical example is shown in Figure 2-3.

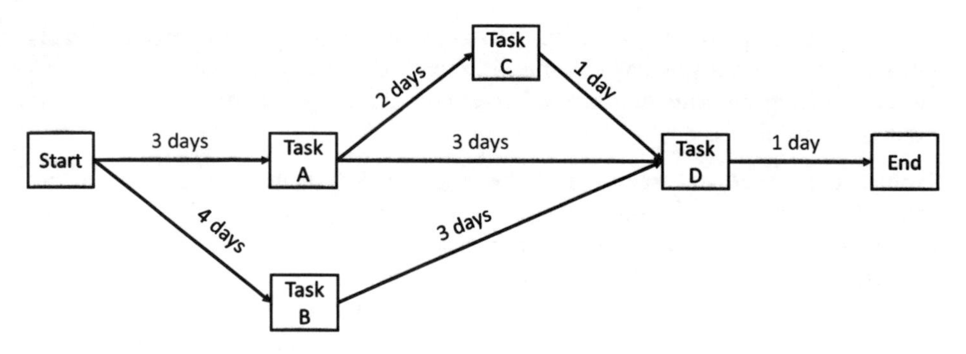

Figure 2-3. Sample PERT chart

This simplified PERT chart shown in Figure 2-3 shows the implication of two options on the timeline of a hypothetical project. The Options shown will consume additional resources of human and material capital. The logic of the PERT chart is that it allows enough time so that resources can be applied to "catch up" to the rate of the overall timeline. Although this figure is simplified, it can apply to such complex issues as submarine or weapons design and implementation. Those models require significant computational power to excise their full capacity. The additional planning capability it offers well outweighs the cost of implementation.

The PERT chart helps delineate tasks in an orderly manner and more importantly showed the interaction of tasks that would allow prediction of the overall project completion.

In the same time frame, the commercial version of PERT was released. It is called the Critical Path Method (CPM). The use of these techniques produced questionable results and was very complex to use. Their lifecycle has been extended as ordained by the Navy's requirement that they be used by the procurement vendors.

In the 1910s, an inventor named Henry Gantt created a bar graph version of a project-planning tool. It featured time-based graphics showing the time to complete tasks and the relation and dependencies of one task to another. There is also a budget-tracking element that allows monetary quantification of the tasks. Unlike the PERT chart, the Gantt chart has survived as a useful tool because of its ease of use and concise graphical presentation. A graphical sample of a Gantt chart is shown in Figure 2-4.

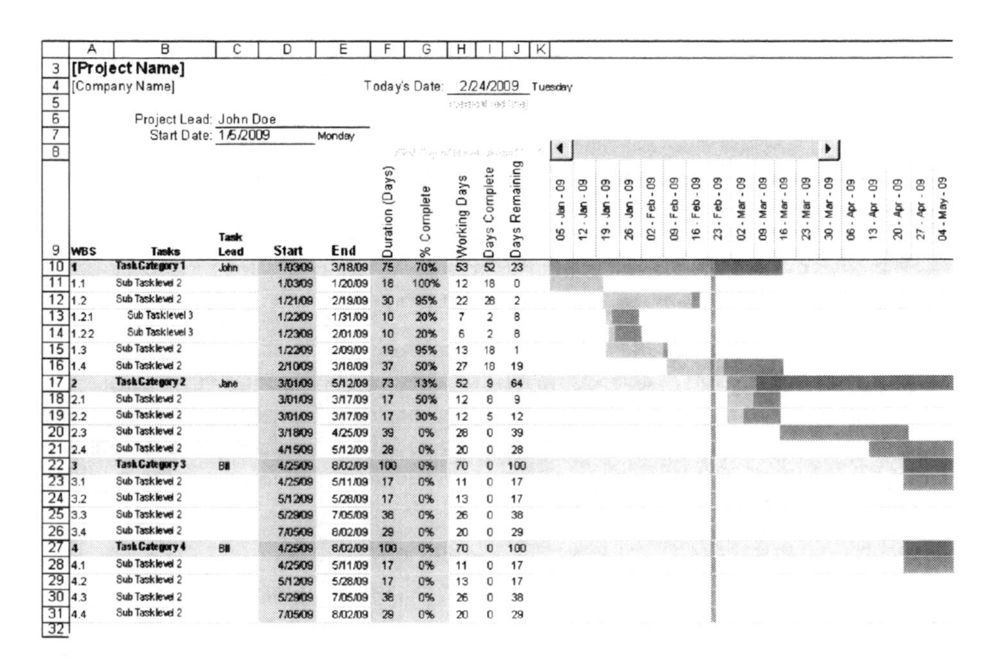

[Project Name]
[Company Name]
Today's Date: 2/24/2009 Tuesday
Project Lead: John Doe
Start Date: 1/5/2009 Monday

WBS	Tasks	Task Lead	Start	End	Duration (Days)	% Complete	Working Days	Days Complete	Days Remaining
1	**Task Category 1**	John	1/03/09	3/18/09	75	70%	53	52	23
1.1	Sub Tasklevel 2		1/03/09	1/20/09	18	100%	12	18	0
1.2	Sub Tasklevel 2		1/21/09	2/19/09	30	95%	22	28	2
1.2.1	Sub Tasklevel 3		1/22/09	1/31/09	10	20%	7	2	8
1.2.2	Sub Tasklevel 3		1/23/09	2/01/09	10	20%	6	2	8
1.3	Sub Tasklevel 2		1/22/09	2/09/09	19	95%	13	18	1
1.4	Sub Tasklevel 2		2/10/09	3/18/09	37	50%	27	18	19
2	**Task Category 2**	Jane	3/01/09	5/12/09	73	13%	52	9	64
2.1	Sub Tasklevel 2		3/01/09	3/17/09	17	50%	12	8	9
2.2	Sub Tasklevel 2		3/01/09	3/17/09	17	30%	12	5	12
2.3	Sub Tasklevel 2		3/18/09	4/25/09	39	0%	28	0	39
2.4	Sub Tasklevel 2		4/15/09	5/12/09	28	0%	20	0	28
3	**Task Category 3**	Bill	4/25/09	8/02/09	100	0%	70	0	100
3.1	Sub Tasklevel 2		4/25/09	5/11/09	17	0%	11	0	17
3.2	Sub Tasklevel 2		5/12/09	5/28/09	17	0%	13	0	17
3.3	Sub Tasklevel 2		5/29/09	7/05/09	38	0%	26	0	38
3.4	Sub Tasklevel 2		7/05/09	8/02/09	29	0%	20	0	29
4	**Task Category 4**	Bill	4/25/09	8/02/09	100	0%	70	0	100
4.1	Sub Tasklevel 2		4/25/09	5/11/09	17	0%	11	0	17
4.2	Sub Tasklevel 2		5/12/09	5/28/09	17	0%	13	0	17
4.3	Sub Tasklevel 2		5/29/09	7/05/09	38	0%	26	0	38
4.4	Sub Tasklevel 2		7/05/09	8/02/09	29	0%	20	0	29

Figure 2-4. Sample Gantt chart

One further twist in the evolution of project-planning tools came with the popularity of computers and cloud-based technology. These developments led to real-time collaborative applications of the tools among multiple players. Online variations of Gantt charts are offered by LightHouse, SpringLamp, Jumpchart, Basecamp, and now Microsoft in their Office suites.

Measurement of performance includes statistical quantification of time and utilization of money. In the latter category, an example that is favorite of mine is Net Present Value (NPV). NPV is a formula that allows the user to bring various timelines to a present time value in terms of dollars. It works best in projects that tend to be more stable and when conditions such as interest rates and the cost of money don't fluctuate too frequently. With project capital allocations now brought to a common time, it becomes possible to compare the performance of multiple projects' performance and to better estimate which makes the most sense financially. In Chapter 6 we will explore the importance of planning and its use as a means of allocating resources to ensure successful commercial outcomes.

Does this Relate to Commercialization?—A Summary

In this text, I argue that a "model" of a commercialization process can improve the probability of success for the outcomes of projects considered by a process. At minimum, it allows for an orderly allocation of resources and establishes metrics for monitoring progress and that can lead to successful outcomes.

Is it absolute? Does it offer a process cookbook? Is it ubiquitous to all projects? Of course not. It does, however, offer a tool for collaboration and the coordination of people, space, and capital. Sometimes I suspect that the major advantage of developing a model is the dialogue that ensues among the participants as they attempt to envision the multiple steps, skills required, and common goals to prevail.

In 1972, I had the opportunity to be on panel discussion sponsored by the MIT Venture Forum in New York city on the role of venture capital in new ventures. I happened to be sitting next to the now deceased Bob Noyce, who was one of the founders of Intel (and an MIT graduate). At that point, the pioneering chip maker company was starting to grow. I asked him if knowing what he did at the time we met, would he have started the company. His immediate answer was, "Certainly not. One of our greatest assets was our naiveté. We didn't know what we couldn't do."

There are varying degrees of uncertainty in the journey throughout the various steps of bringing ideas to commercial reality. One clear advantage of an orderly progression is that the uncertainties and the assumptions that surround them can be quantified in a logical and repetitive manner. That alone helps bring order to a complex and uncertain system. It also carries the risk that in something like commercialization model development, which requires a degree of creative innovation and intuition, we might lose those elements to a more rigorous process. But as you'll see, disciplined program reviews—where the progress and creativity of the project's progress can periodically be examined and of course altered to meet changing external conditions—can mitigate against that danger.

CHAPTER

3

The Sources of Ideas for Commercialization: Are There Enough?

The Limits of Myopia

In the first chapter, I suggested that there are an abundance of ideas, projects, and even companies to fulfill our global needs for innovation and entrepreneurship. Yet, the economic measurements of employment, financial growth, and wealth distribution are lagging behind this idea. What might be needed is a broader, more dynamic perspective in how we identify what those ideas and opportunities might be. Let's look at some of the ways you find them.

Sources of Ideas

Finding ideas worthy of becoming projects for successful commercialization is a critical part of the entrepreneurial journey. As suggested in the last chapter, it isn't that there are not enough ideas. Rather, there is an abundance of possible avenues for achieving the project's goals. This is where the process begins. It is to ensure that as many viable sources as possible are considered for the overall model to be as effective as it can.

By sorting out multiple sources of innovation, it's important to avoid repeating blindly the use of singular source channels. Actively and dynamically seeking out the best match of source to a given project is critical. In Figure 2-1, you saw five traditional groups of resources for technology commercialization. It is not meant to be an inclusive list, but rather a partial selection of alternatives.

As noted in Chapter 2, aircraft pilots who are certified to fly airplanes in metrological conditions that lack outside visual references (such as when flying through cloud banks or heavy traffic) follow Instrument Flight Rules (IFR). In the training for that certification, pilots learn to use six primary instruments to fly the aircraft. Training includes the skillset of visually scanning the instruments for information. If the dynamic scan is broken and the operator becomes fixated on any one instrument, it becomes a recipe for failure.

This parallels the quest for a similar process that's utilized in seeking business opportunities quite well. Fixation on any one path invites failure, and yet we see many examples of a rigid direction early in the project.

It is tempting to look to invention first. Much has been written about creativity and role of developing fundamental ideas.

Yet in-house invention is perhaps the least likely avenue to success because of the many unknowns and risks. That's why Figure 2-1's focus is on multiple sources and why it starts with looking at licensed existing technology and its inherent strengths of being technology already reduced to practice.

Licensing

Licensing as a source of opportunity begins with understanding how patent law works. Governments create a basis for protecting inventors' intellectual rights to their ideas by issuing "letters of patent" (or a copyright for written materials). Either establishes a limited monopoly of use by the inventor. The theory is that the nature of invention is so fragile in its inception that it takes a period of time to realize the invention's value to society. If this one element of risk can be mitigated by the government issuing patents, then the overall economy benefits and invention is encouraged. Within the realm of patent protection is the right to allow (license) others to use the patent's domain. This is a powerful means of exploiting the patent's value when the individual(s) to whom the patent is issued do not have the means or interest to commercialize it.

Before the merits of licensing are explored, it is best to look at the fundamentals of patents. This will help you understand the value of licensing as a strategy of sourcing ideas to commercialize.

Patent Law

A patent is an exclusive right issued by the federal government to an inventor(s) who claims a particular field of creative invention. It says that no one can infringe or copy the inventive fields claimed in the patent for a term of 20 years (in the United States). It is a monopoly for the use of the idea for that period.

Knowing where you are in that time period is relevant. More recently issued patents have the longer market potential for yielding value. Mature issuances are just the opposite. This particular aspect is considered by investors who attempt to establish a financial value of the IP portfolio when performing a due diligence analysis of a potential investment opportunity. They use a model of an "aged timeline" in their calculations.

The form of the issued patent is quite formal. It includes:

- Abstract of the invention
- Formal drawing
- Field and background of the invention
- Detailed description in which the claims of the patent's scope are delineated

Conventional wisdom is that crafting a patent application is in the domain of attorneys. That is probably good information. This particular patent was issued to me while I was at Phoenix Controls (see Appendix A). Note the detail of the claims section. It is here where the scope of the patent is defined. Also, it is interesting to see how much information is revealed to the public. That is the tradeoff of the security gained.

There are privileges and responsibilities that go with the issuance of a patent. The first consideration is how formal and broad the protection offered by the patent is. If the claims are too broad, the patent becomes vulnerable to loss of novelty and rejection during the application process

The entry level of the journey to intellectual property protection is the provisional patent. Written informally, it gives the inventor time to establish the case for developing the invention/idea further while declaring the date (time) of the application. Its term is only one year. It is a simple exposition of an idea. Sometimes it is a copy of a scientific journal description or similar non-legal terms. It clearly describes the enabling technology but defers the argument or claims of inventions. That is left to the more formal utility patent. Expenses are not trivial. The fees for a provisional patent are about $1,500 to $2,500 to file with a lawyer. You can file your own provisional patent for less. The tradeoff of the provisional patent is that it is vulnerable to copying and alteration of the original idea.

The next step in securing the formality of patent rights occurs within one year of the filing of the utility patent (if filed). The utility patent is more structured and better defines the protection it enables. It is also a more elaborate and expensive process. You might pay $25,000 or more for this category of protection. The process is best accomplished with a registered patent attorney. The application process may take several years to complete. The final issued document is a structured device that has detailed drawings, a description of the patent's intellectual property, and even more important, a listing of the fields of invention you are claiming.

This litany of the various aspects of the inventions is what describes the patent to the world. The detail and scope of the claims of the patent are the value added by a good attorney. If the claims are too narrow (or too broad), they weaken the area of intellectual property the patent is trying to protect.

The application process also tests the idea's novelty (that is, that no prior art or publications exist) and whether the idea can be reduced to practice by someone knowledgeable in the art. Once the patent is granted, a number is issued and its importance to the commercial cycle can be examined. Traditionally, patents centered on fundamental inventions. Today, the field embraces biotechnology, software, and even aspects of the human genome. Where these will lead and their implications are uncertain.

As you are probably aware, patents must be defended. If others have infringed on the space defined in the patent claims, then they must first be put on notice. If they are noncompliant and do not desist the violation or infringement, they must be challenged in court to resolve the interference. It something like the eminent domain rule of real property whereby if you allow someone to walk across a section of your land and do not challenge it, you lose the right to deny use of the property. Failure to contest infringement may even invalidate the patent. Patents also require that you pay annual maintenance service fees to the government.

What has been described so far is the situation in the United States. But we live in a world of global competition and collaboration, and so there is need for establishing patent rights internationally. In 1970, a global conference was established in Paris to allow cross-filing of patents in multiple countries. It wrote the Patent Cooperation Treaty (PCT), which established rules and procedures for mutual filings around the world. It is now administered by the World Intellectual Property Organization (WIPO), located in Geneva, Switzerland. Under the PCT, patents are individually cross-filed in select countries where future business is anticipated. The patent can be amended within a period of 18 months to include additional countries. The cost of a PCT filing is significant. It can be in the range of $100,000. There are additional ongoing maintenance fees in each country.

Once IP is secured, an intriguing discussion evolves about the strategy and use of it as an asset. Some companies file for patent protection as a defensive strategy and accumulate large portfolios. Others file only when it is critical to specific applications and thus have lean portfolios. No one universal strategy has evolved.

One would hope that there was a nice, orderly global system for the protection of ideas. Unfortunately, the commercial world has not evolved that way. In certain emerging countries, like China and India, it is quite common for individuals and corporations to ignore the due process of U.S. and European patent law and challenge (and infringe) the intellectual property's validity on a global basis. The future direction of those challenges is quite uncertain. In China, for example, there recently seems to be an increased awareness of the value of IP. This is particularly true in academic research sectors.

Patents are not limited to the utility patents just described. There are other broad categories of defensible IPs in the design and natural plant areas. There are specific patent formats for these. Also there are trademark and copyright categories where variations of protection are made over a period of time.

A Proprietary Alternative to Patents

There is an alternative to nurturing inventions under the protection of patent law. It is a strategic one that inventors or their organizations can utilize. It involves holding the technology (or idea) as an internal secret. The concept is called ***proprietary information***. One penalty of the patent application process is the need to openly divulge an invention's workings to public scrutiny in great detail. Securing the ideas as proprietary information provides the alternative of not doing so.

This alternative requires significant internal discipline and the possibility of error because of the secretive nature of withholding information. A great example is the Polaroid Corporation. This now-defunct pioneer of instant pictures created in a chemical process in the photo's packaging that utilized proprietary information in a unique manner.

The company was co-founded by Dr. Edwin Land, a scientist who resided in Cambridge, Massachusetts. At its peak, Polaroid employed 16,000 employees in its R&D laboratories. Of them, only three trusted employees knew the overall chemical process, and the rest worked on the components. That is an astonishing ratio!

The magnitude of the use of proprietary information is not well documented because of the very secretive nature of its application.

IP and Innovation Commercialization

In 1980, the U.S. Congress passed the Bayh-Dole Act. Its impact was significant in that it allowed universities that receive federal research funding to "own" the resulting intellectual property (IP) that came from the funded research and thus receive a monetary benefit from the revenue generated by the license fees it creates. The Act was considered win-win legislation in that the government could benefit from the commercial use of the R&D it funds, and the universities could realize a new source of revenue. For those looking for creative ideas, it proved to be a robust new source.

Universities responded by creating technology licensing offices (TLOs) to promote user access to their research. It has become big business, with annual revenues in the hundreds of millions of dollars. MIT's Technology Licensing Office, for example, is staffed by over 20 professionals. So what are the implications? In simple terms, it has accelerated the movement of science to the marketplace and fostered competitive economic benefits for the academic research once cloistered in the university labs.

Certainly, the effect of patent law protection reaches beyond universities to inventers and corporations. In companies, it has additional strategic value in what is referred to as ***in-licensing***. That the owner of a patent can license others to use his or her technology is understood. With in-licensing, that success is defined by being integrated with other licensed IPs, yielding a combined effect. A new or innovative idea may require other technologies to complete it. Often it is the basis for collaboration and joint ventures. In early venture investments, funders assess the strength of the patent portfolio as an element in their valuation calculations. In certain cases, in-licensing allows early-stage companies to jump start their business plans.

An example of this is seen in a company called Phoenix Controls in Newton, Mass. It is a company I co-founded. To accelerate our entry into the market, we licensed a critical piece of technology from MIT that enabled us to start selling product to a limited segment of our customer base in our first year. The full system approach was not ready for several years. Otherwise, the internal R&D team would have taken several years to become effective in developing its own product solutions. Paradoxically, when the company's equity was liquidated to a Fortune 500 company 10 years later, it was observed that early in-licensed technology contributed less than 5% of the company's cumulative revenues. The value of the quick start, for market entry and brand recognition, was immeasurable.

Tip Licensing others existing technology can in some cases help you introduce products to the market much quicker than you might otherwise.

In the case of university-licensed technology, R&D is funded by the government and is often accomplished by world-class research teams. Indeed, that is a plus. Some argue, however, that if the R&D is accomplished in internal company laboratories, it is more focused and can contain proprietary information. This debate of outside vs. inside is a dynamic and continuing one. Research lab managers are reluctant to give up resources while product-development folks are eager to move ideas to market. It is usually resolved by the strategic needs of particular projects.

Is there a downside to licensing? Of course. One obvious negative is that the theory of the patent's invention and its method of reduction to practice are explicitly delineated in public documents. It can be argued that a competitive inventor can thus see the shortfalls of the declared invention and begin to invent an even better solution. It becomes a risk/reward strategic situation.

In addition, the learning curve of discovery (and mistakes) is held by the developing research team. This wisdom can become critical in developing follow-up products.

A second important consideration is the cost of the make/buy decisions regarding the patent license. Certainly, corporate funding of research is a significant expense. There is a continuing decline in attractiveness of this alternative. On the other hand, license fees can be significant as well. In addition to royalty fees, which are ongoing and absorb a percentage of the unit's gross margin profits, there are both upfront payments and penalty fees that must be considered. They go directly to the company's balance sheet and income statements. It's best to warm up the accountant's green eyeshades and sharpen their pencils! The cost/benefit must be carefully calculated and is best resolved in negotiating the license agreement.

There is an organization that gathers university and research organizations' IP operating data and best practices. It is called the Association of University Technology Managers (AUTM). Each year, it performs a survey of their members and publishes the resulting aggregate data in an annual report. In the Fiscal Year 2013 annual report it noted that:

- 719 new commercial products were introduced
- 5,198 licenses were executed

- 1,356 options were executed
- 469 executed licenses contained equity as well as financial fees
- There were 43, 295 active licenses and options
- 818 new companies were formed from the issuance of university-based licenses, 611 of which had their primary place of business in the licensing institution's home state
- 4, 207 startup companies were formed from university-based licensing and are still operating as of the end of fiscal year 2013

Still to be decided is whether this line of technology commercialization is appropriate to a given project. This will be discussed later in this book. Whatever the path, the magnitude of operational data from this source certainly argues for close consideration.

Professor Alexandra Zaby of the Tubingen University business school in Tubingen, Germany (Springer Verlag Berlin 2010) performed a detailed analysis of the topic for her book *Decision to Patent* and confirmed the marginal utility of the decision. According to Zaby, the costs and disclosure of the information may outweigh the advantages of the patent.

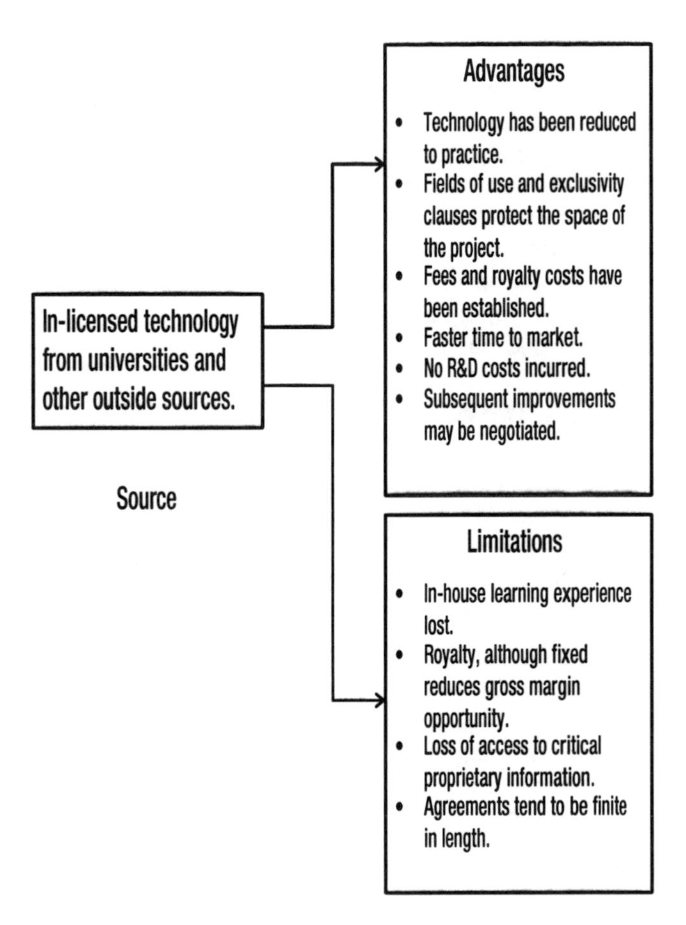

Figure 3-1. In-licensing of intellectual property

Operational Questions about IP

Securing access to IP is an intriguing alternative. But is it the right path for successful commercialization? Let's look at some fundamental questions:

- Is there an inherent advantage such as time to market, market share, broader product offerings, and so on, as a basis to securing rights to a patent?
- Is there sufficient pricing marginality to support the additional financial burden of licensing fees and royalty payments?
- Would the risk of possible litigation be strong enough to move forward without IP? Lawyers refer to this as the "freedom to operate" and it points to the need to analyze whether the surrounding IP is too mature or weak enough in its competing claims to support the risk.

Licensing becomes a dynamic source of innovative technology for emerging project and new companies. In many cases the technology secured by IP may have been developed at a time and place that is no longer relevant. This is certainly true of university-based research, where the incentive to directly commercialize it is beyond the purview of the halls of academia. In certain sectors (such as biotech or nanotech), where the development cycles are long and expensive, the idea of transferring it to an emerging entity, such as a division of a large corporation (e.g. a major pharmaceutical company) becomes attractive. Internal R&D offers an alternative. It will be interesting to see how it converts to licensed ideas.

R&D: Beyond IP Strategy

Internal corporate R&D has been a cyclic and powerful resource for innovative technological change. In the 1950s, the landscape of R&D facilities seemed both lavish in design and endless. Architectural awards were granted to new, state-of-the-art research facilities. These labs were part of the draw to attract top talent. They were always featured in corporate annual reports.

Today, many of the dominant players no longer exist. Homer Research Labs at Bethlehem Steel, Sarnoff Labs at RCA, Bell Labs at AT&T, and PARC at Xerox all are examples of world-class research facilities that have been either decimated or closed. What happened? More important, what are today's trends to replace their function? And of course, are R&D labs a viable source of ideas for technology commercialization?

According to Booz&Co.'s *2014 Global Innovation 1000* report,[1] the aggregate R&D spending in the United States is $637 billion. The number is not precise and includes government research facilities (minus classified projects) and not-for-profit organizations. As large as this seems, it is not the largest amount internationally as a percentage of GDP. Israel, for example spends 4.2% of its GDP on R&D, compared to the United States at 2.6%, according to a Battelle 2014 Global Funding Forecast.

There are other interesting notes about R&D spending. They include:

- The 11-year compound annual growth rate of R&D spending now equals 1.4%. The three-year rate is 9.5% (Booz&Co. report). The decline is significant.
- Three industries dominate the spending. They include computing and electronics, health, and automotive. These three sectors comprise 65% of all spending (Booz&Co. report).

[1]www.booz.com/media/file/BoozCo_The-2012-Global-Innovation-1000-Media-Report.pdf.

- China and India are growing faster in R&D spending per capita than mature regions. Combined, they are growing at a 27.2% annual rate while the average for all other countries is 9.6% (Booz&Co. report).
- Sales and R&D spending of certain sectors and companies correlate to sales growth. However, R&D as a percentage of sales has been in steady decline for almost 10 years. It is now an average of 1.14% (Booz&Co. report).

This is a fairly dismal look at the U.S. R&D effort. One possibility is that alternative means of innovative development are being utilized and thus reliance on the classic R&D function is in decline. On a global basis it is clear that China is increasing its per capita R&D expenditures.

In September 2000, the National Institute of Standards and Technology (NIST) asked the National Research Council to assemble a committee to study trends in science and technology (S&T), industrial management, and how research and development would impact the introduction of technological innovation. The results were published in a report entitled "Future R&D Environments" (National Academy Press, 2001). As part of their findings, the council observed the accelerating change of the environment in which the research was being conducted. They noted that all of the following were changing the nature of the work—a move to decentralize research and conduct it at different locations in the world, the outsourcing of projects, new issues regarding privacy, the advent of biotech and genomic research, and finally the declining span of product lifecycles.

In the midst of this, the Industrial Research Institute in its 2013 R&D Trends Forecast found that licensing fees and hiring graduates seemed relatively flat, yet stable.

R&D Strategy

Putting this all together, it now becomes important to look at R&D as a source of ideas and innovations in the path to commercialization. In all of the studies cited, there appears to be a positive correlation between R&D and revenue. High fliers like Apple, Intel, and 3M are examples with expenditures of R&D as a percentage of revenue in the low teens. Mercedes Benz and other automotive companies were shown in the middle of the list but now show increases as they try to move to new technologies.

When financial and investor professionals look at income sheet expenditures in functional areas such as marketing and R&D, they coined a phrase called "comparables" (or "comps" for short). Comps relate expenditures of different categories in other companies with similar market or SIC sectors. Comps essentially argue that if you want to be in the fast-moving electronics business, for example, you must be prepared to allocate 8–12% of the corporate revenues to R&D. National averages of corporate research are closer to 1–2%. Were it that simple, it would be just a matter of upping the numbers of dollars invested.

Andy Grove was one of the founders of Intel. The company manufactured computer memories and operational microchips. Repeatedly, it was the leader in the manufacture of computer memories and later of processor chips. Andy was publicly quoted as saying that, "Success breeds complacency. Complacency breeds failure. Only the paranoid survive." Ted Leavitt, the noted Harvard Business School professor who wrote the classic Harvard Business School case entitled "Marketing Myopia," chided established businesses like the railroads to understand "what business they were in." Railroads, argues Leavitt, failed because they thought they were in the railroad industry, not the transportation business.

Internally, we see that corporate culture and clarity of vision are needed to guide allocations of resources for the future in both R&D and capital improvements. Today, a new layer of constraints in regulatory issues, global competitiveness, and availability of material resources challenge research to be more nimble and more adaptive while certainly being more innovative.

As the economy moves to more science and technology-based platforms like biotech, nanotech, cleantech, and so on, the pressure to create internal R&D capabilities is reappearing. Look at Kendall Square in Cambridge, Massachusetts, for example. It features block after block of technology-based companies with almost incomprehensible biotech names. Certainly the proximity to MIT's research labs impacts this approach. Some of these startup companies are doing very well indeed. An example is Novartis, which is currently going through a third major expansion of its Cambridge research facilities on Massachusetts Avenue.

Academics have certainly argued the case for and against internal R&D efforts. In an article published in *Research Policy*[2] entitled, "Internal R&D Expenditures and External Technology Sourcing," Reinhilde Veugelers examines multiple variables such as size, diversity of projects, and impact of external sourcing on performance. He concludes that the real advantage of internal R&D is that it allows "absorptive capacity" to embrace new ideas both internally and through external collaboration.

[2]Vol. 26, Issue 3, October 1997, pp. 303–315.

In a 2009 Harvard Business Review case entitled "Merck (in 2009): Open for Innovation?" (MH 00009), the authors Alicia Horbaczewski and Frank Rothaeramel continued to explore Merck's ability to innovate and stay competitive by reaching out to universities, research institutes, and other company collaborations. They observe that Merck's total R&D expenditure was only 1% of the world's effort in fields that were emerging rapidly but were too complicated for Merck to navigate alone.

In the March 2007 issue of *The Economist*, an article entitled "Out of the Dusty Labs" traces the development of the Corporate R&D labs to Vanover Bush (Franklin Roosevelt's science adviser and later president of MIT). He published a paper entitled "Science, the Endless Frontier" in which he envisioned a cooperation between government-funded research in the universities, corporate labs, and the military. Indeed, all that happened.

After World War II, it seemed that big R&D labs became a luxury that only a few high-margin companies (like IBM and RCA) could afford. Yet, Noble prizes for the transistor and laser (AT&T) and for the mouse and computer graphical interface (Xerox) came from those labs. As the technology of information entered the arena, the larger vertically integrated labs no longer were required. Eric Schmidt, founder of Google, was quoted in the same article as saying, "[the] smart people on the hill" method no longer works. Instead, researchers have become intellectual mercenaries for product teams. They solve immediate needs.

Older pundits who poked at the R&D function used to say it was isolated by attitudes like NIH (Not Invented Here) and throwing ideas "over the wall." (Note: One exception is IBM, where the research folks actually carry their ideas to manufacturing.)

Again, in *The Economist* article, John Seely Brown, the former director at Xerox-PARC is quoted as follows, "When I started running PARC, I thought that 99% of the work was creating innovation and then throwing it over the transom to dumb marketers to figure out how to market them. Now, I realize that it takes at least as much creativity to find ways of bringing ideas to reality. Knowing that, I would have spent my time differently if I had figured this out earlier."

And people wondered what happened to the dinosaurs—PARC no longer exists! Perhaps corporate R&D will not be a leading candidate for commercial opportunities.

R&D: Question the Viability

All of this leads to a series of questions about a given organization's R&D effort. For example:

- Are measurable R&D metrics such as industry- compatible expenditures, number of patents, number of peer reviewed articles, number of new products produced, and so forth, on a par with other organizations in their field?
- Is the evidence that the R&D function is "open for innovation" and has the absorptive capacity to utilize both internal and external sources of ideas?
- Is there a close, working organizational connection to marketing and corporate vision within the R&D function?
- Are long-term and short-term corporate needs for innovation, new ideas, and market competitiveness being met by the R&D function? Has there been a consistent funding pattern of R&D that transcends cyclical trends in profits and cash flow?

There are more alternatives to commercialization; it reaches beyond formal R&D and licensing others' ideas. Let's continue our search for alternatives.

Extending the Product or Market

Marketing and product design professionals have long known the value of extending the life of product or service design by using product and market extenders. In product lifecycles (see Figure 3-2), there is an inevitable decline in the products' acceptance or market dominance. This can be caused by technology obsolescence, competitive forces, or simply newer and better ideas. There are both good and bad reasons for allowing the inevitable decay of the end-of-life portion of the curve.

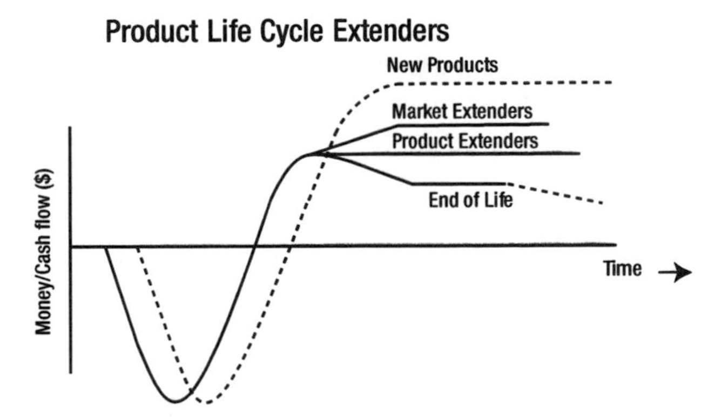

Figure 3-2. Product lifecycle

If inevitable, there are distinct end-of-life strategies available. For example, retailers use clearance sales on a regular basis. It is those companies who do nothing that are at greatest risk of losses. On the plus side are simple economics of technologies that have significant frontend or tooling costs that are simply not cost-effective to change. In the cases where there is both a lack of nimbleness or attention to the need for change (lethargy) and a lack of market awareness, the probability of losses increases.

In either case, left unattended, the product fails. If this weren't explicit enough, the concept can also be extended to corporate models—some organizations simply become irrelevant.

Can the severity of the inevitable product declines and incremental change be reduced? Of course. The most dramatic example is in the computer chip manufacturing cycle. A toolset, or "die" as it's called, is used to produce multilayer computer chips. They are estimated to cost in the billions to fabricate. Clever designs actually build a family of products into a single chip. Business and market conditions determine when to enable features in a timely way that allows the products to have a new life. In that sense, they mitigate the end-of-life phenomena.

Hard good appliances and automotive designs sometimes use simple sheet metal alternations to imply that they are "new" products. Usually these changes are planned at the beginning of the product or service inception.

Perhaps the most prominent example of this process is the evolution of Microsoft Office. Certainly an older product, it has reinvented its feature set six times to its current offering as Office 2013. Each release offers new and extensive features. In a diabolical manner, the companies will not support over two earlier versions, thus ensuring demand for newer versions and upgrades. It is estimated that the true cost of the software is about $1,000 for the user to maintain its currency.

Although the powerful Office product offers thousands of features, the majority of users use only a small number of the functional attributes. Microsoft also saved the billions of research dollars (and time) required for new product launch. Today, the product's dominance is being challenged by open software that seems fully featured.

Engineers coined a term called "feature creep." It is enticing to endlessly add features to a product as it reaches its maturity. Certainly, it helps generate additional revenue. Like the little "I" of incremental innovation, it also draws resources from the need to innovate and change the environment of the product.

This issue is not constrained to software. An interesting example is the Sony Walkman. At its inception, it pioneered how we listen to CDs. Controls were simple and rugged. As the product matured, it became "jogger shock proof," "waterproof in the rain," and had auto-reverse features added in an attempt to differentiate the product. While doing all this, Sony lost sight of the fact that the CD had already reached its peak and that CD use was declining with respect to other forms of music transmission. What were they thinking?

Each path of the innovation journey has its plusses and minuses. Product (and market) extenders allow us to weigh the incremental gain (and incremental expenditures) against product obsolescence. That balance must also be compared to the risk of delaying disruptive and innovative change, a significant risk. Certain ideas will also be of benefit. It should be the strategy to ensure that they be added to the possible sources of commercial projects.

Some Questioning Thoughts: Product Extension Food for Thought

Consider the following when you view the opportunity to extend a product's life in some manner:

- The product (or corporate) lifecycle predicts an inevitable end of life. Is there a plan for managing that process? Are you considering it? Are there anticipated options?
- Is there a way to utilize the mature products to your advantage?
- Does the product design incorporate a "platform" philosophy that anticipates families of products?
- Are the market cycles coincident with the product (or service) cycles?
- Can the product and service offerings produce adequate financial returns, or are they increasingly incremental?

Joining and Buying: The Role of Mergers and Acquisitions (M&As)

Venture capital as we know it has its roots in a model propagated by Georges Doriot in his launch of a venture fund called American Research and Development (ARD) in the 1960s. He preached the need for measured risk financing and "nurturing" early-stage companies. The concept was to appeal to large pension funds to become limited partners in ARD. The funds were conservative in their nature and reserved small percentages for high risk/high return investments to boost their portfolio's performance. The Venture Capital Fund model serves this need well.

The concept relied on liquidation of the venture fund's investments through an initial public offering (IPO) of the equity (stock) in their portfolio companies. The highest returns (10–20 times) were available through IPOs. IPOs, however, rely on an available marketplace for new issues. This market, in turn, fluctuates with stock exchange performance and is not always available to accept the IPO offerings. Indeed, this was the trend for several years after the recession of 2008.

So what happens to the investor's need to liquidate their investments? They simply revert to another strategy, which is to encourage a company to buy another company in their portfolio (acquisition) or to merge with another in exchange of stock (merger). Neither of these alternatives has the potential returns of an IPO but are still quite attractive. In this category is the strategic sale, where there is a close alignment of market or technology interests. They have the highest premium.

The M&A Alternative: Questions

When you are considering a merger or an acquisition, find suitable answers to these questions:

- Is there a clear and specific objective to the M&A transaction beyond capital growth?
- Is there a post-transaction implementation plan? (Most M&A transactions unravel after the deal without a clear plan).

- The M&A transaction alters the balance sheet of both companies and raises the question as to whether the "newco" is strong enough financially to grow. This dilution is often mitigated by debt in what is referred to as a "leveraged" buyout. Is this good for the new entity?
- Possibly the most complicated issue of the post transaction is the alignment of the succeeding organizations. Overlapping jobs are eliminated. Good planning can mitigate the human carnage that goes with it. Is there a pro forma organizational model for the "newco"?

M&A emerges as an obvious strategy for growth and innovation. The basis for this activity can be strategic and allows options for new technology and market developments. The economic considerations flow from a belief that the target is undervalued and thus ripe for the transaction. Sometimes the motivation is less than altruistic and may be driven by ego or simply hubris. Investors see a decline in IPO activity and see an M&A as an alternative basis for liquefying their investment.

Invention/"Other"

This category of sources for commercializable ideas is possibly the hardest to define. Ideas that form inventions exist at the intersection of recognized problems, creativity, and irrational thought. It is characterized by an "a-ha" moment when useful ideas are generated. Thomas Edison said that, "There is always a way to do it better ... find it."

In a book entitled *Innovate Like Edison*, authors Michael Gelb and Sarah Miller Caldicott cite other comments Edison made. They include:

- "My philosophy of life is to work to bring out the secrets of nature and apply them for the happiness of man. I know no better service to render during the short time we are in this world."
- "I start where the last man left off."
- "To invent, you need a good imagination and a pile of junk."
- "To have a great idea, have a lot of them."

Edison went a bit further and proposed an additional five Competencies of Invention. They include:

- Solution-centered mindset
- Kaleidoscopic thinking

- Full-spectrum engagement
- Mastermind collaboration
- Super-value creation

There is significant explanation in the literature about Edisonian thinking. To suggest that it might be codified and set into rules is of great interest. Certainly, organizations should attempt to capitalize on it. Inventors have been characterized as loners and tinkerers. But sweeping, broad-brush descriptions don't help to understand inventors as a group; it really doesn't get you any closer to understanding how to embrace, extend, and capitalize on their skills and creativity.

There are multiple organizations to help those inventors who are willing to participate in group activities. In Springfield, Massachusetts, for example, there is a group called the Innovators Resource Network.[3] They meet monthly in a local technical school and discuss a wide range of topics ranging from patent protection to new technological processes for making prototypes. It is a very applied and tactical group. Its membership, and alternate meetings, are closed to the public.

At the national level, there are organizations designed to assist both experienced inventors and those just starting out. The National Inventors Hall of Fame, located in the U.S. Patent and Trademark Office in Alexandria, Virginia, is an example. It showcases many ideas and projects to inspire and motivate invention. The organization sponsors meetings and workshops and provides informational resources. They also sponsor a camp to help younger inventors polish their skills.

Nurturing the Inventive Process

From federal incentive programs to state-level assistance and invention incubators as well as private incentive groups, there are numerous resources of space and mentor support available to inventors. Invention is considered the lifeblood of a vital economy and a source of a country's ability to compete. Repressed societies are characterized by the country's being noncompetitive and not economically viable.

There is a creative element to invention itself. It is that invention is almost like an art form of creativity. It differs from the innovative skills discussed in this text and so many current articles. At least one element of the inventive process is that inventors seem to require elements of solitude or at least distance from the conventional organizational structures.

---

[3]www.irnetwork.org/.

When I worked for a company called Waters Associates in Milford, Massachusetts, there was very creative individual who in many ways was a disruptive element in the development of new products. He would challenge each step and attempt to change things at the last moment on a seeming whim. Yet, his ability to conceive bold changes in the core business model was, in the parlance of the kids, awesome! The eventual resolution of this conflict was to create a "skunk works" five miles from the central plant that was equipped with machine tools, small labs, and generally enough space to create next generation concepts. On the one hand, this was an expensive and organizationally awkward solution, but on the other, it allowed creative processes to flourish.

Dean Kamen, the prolific inventor and engineer, goes further. He believes the root of the creative process is created in the early education years. He said that he failed as a student because he always tried to invent answers to quiz questions that weren't on the exam just to be creative. And thus his answers were judged as "wrong." He attended WPI but left before his graduation. He went on the develop the digital infusion pump for controlled intravenous injections and later, the now famous Segway transporter.

In addition to these major innovations, he focused on early education and created a university/industry collaboration called the First Competition to foster curiosity in engineering and science. It featured robotics teams' solutions to a statement of a given problem. The effort focused on high school grade levels. The impact of this project has been that tens of thousands of students and thousands of industrial collaborators have participated in these annual competitions.

There is an inherent conflict in the corporate model and the nurturing of invention. Within the corporate model, there are incentives for completing projects on time, within budget, and in an orderly manner. Creative invention requires almost the opposite. Some companies, such as Google, allow employees 20% of their time just to innovate and create. They do so within a structured environment that is based on eight "pillars of innovation" that allows for failure, provides café space for discussion, features open accounting-level data, and sharing and information systems that are dynamic and distributed widely in the company.[4]

Throughout the system is a pervasive culture of empowering its employees. Invention, as well as new ideas for products and service improvements, can be generated in multiple ways. Learning to be attentive to the random possibility

[4]*Google Think Insights* in an article entitled "The Eight Pillars of Innovation" by Susan Wojcicki, July 2011.

of commercially viable ideas is important. "Other" represents a placeholder for those ideas not captured in more formal processes. To put inventions in perspective, consider several questions:

- Is the culture surrounding the inventive process encouraged and rewarded, or is the opposite true?
- Are there incentives for risk, thinking out of the box, and even paradigm change?
- Google is certainly an exception in the many ways it encourages innovation. What specific policies in a given organization are there do you have to encourage change and innovation?
- Is there a means for rewarding the change agents? In some companies even the issued patents are posted in a hall of fame.

Invention and "other" new ideas occur at almost any stage of an innovation's entrepreneurship. The challenge seems to be to allow a culture or environment that encourages it and reduces the element of risk. "Other" is that delightful category of ideas that is spontaneous and brings the responsibilities of being open to those ideas that may be "outside the box" and counterintuitive. Those ideas may also be quite valuable.

The World of Ideas: Summary

Wow, we live in a world of many ideas that may come from diverse and unconnected sources. There are different strategies to achieve their commercial success. They can be used in parallel and in combinations. Opportunities can be found in multiple venues and at different stages of maturity. The first lesson of this chapter is that we must remain open to the ways that ideas can be generated and pursued. Like the pilot's instrument scan, the risk of fixating on any one channel is failure. In succeeding chapters, you'll look at how to distill the multiple sources of ideas to the ones that have the highest probability of commercial success.

CHAPTER

4

Winnowing Down: The Challenge of Opportunity Recognition

Recognizing the Opportunity

In the previous chapter, I presented a process model that searched multiple sources of ideas and then distilled them into possible opportunities that are aligned to the overall vision of the environment they were meant to serve. That process can be envisioned as a "funnel" in which ideas are constantly being drawn down to possible candidates.

The frequency and magnitude of the flow of ideas are determined by the capacity of the organization to absorb them. The importance of a "flow" is that not all projects will equally succeed. So the rate (that is, the number and breadth) of idea flow must be considered in a way that a) allows for failure of some projects and b) is robust enough to satisfy the company's (or project's) appetite for new commercial ideas. It is a probabilistic model that relies on a number of ideas being processed at any time. It is not an absolute number. The time to restart the flow of ideas from the beginning is too long to recover in time for the newer projects to be useful. The model is a continuum of ideas to projects to be processed in parallel, not a serial path of evaluating one opportunity after another.

A further importance of the funnel model is that each organization and field of practice has its own dynamics. In biotech, for example, the time to commercialization of project flow is regulated by outside agencies and consumer acceptance. It embraces years of approval processes and trials. Apps and social media projects, on the other hand, have almost instant access to markets. Most funnel models require the dynamic mix of different projects to help smooth out possible disruptions of different time schedules.

As a project's "journey" from sources of ideas to commercial reality moves along, it does so through a series of decision "gates." The gates metaphor is used because it suggests that a conscious decision to move forward is accomplished at multiple stages (or gates). Any gate also allows for a consistent and rational basis to reject a project. A negative decision is as useful as a positive decision to go forward. It can close out inappropriate projects and preserve access to precious resources.

Before we examine the gate concept further, let's take a step back. Before opportunity projects are even considered, examining the broad world of ideas becomes useful as it allows the user to calibrate the landscape of all opportunities available. An "idea" is just that. It is a concept based on intuition, experience, and perceived need—or just a whim. It could be the output of a creative process of searching or even an opportunistic look at markets and available products. Sometimes it needs a curiosity driven "Edisonian" mindset coupled with a need to create, a sensitivity to a customer need, or a confluence of existing ideas.

The use of Global Positioning Systems (GPS) for automotive navigation might be an example of this. Fixed-base and handheld GPS existed for a long time in military and industrial uses before someone decided to apply it to automotive dashboard applications. In most cases, the transition to consumer use contains an element of creative energy and/or entrepreneurial innovation that allows the inventor to see the idea application even before it's real.

How the transference of an idea to opportunity lies in the mindset of a project or a company is a defining aspect of "Opportunity Recognition." Professor Ulrich Kaiser of the Institute of Strategy and Economic University (ISU) of Universtat Zurich in Switzerland offers a course entitled "A Primer in Entrepreneurship," in which he defines an idea as a thought, impression, or notion that might not meet the criteria of an opportunity. He further goes on to say that the transition to an opportunity is driven by business, social, technological, and political forces that help define the product or service opportunity's boundaries. Usually there is an entrepreneur or entrepreneurial environment that fosters the change.

The actual decision to move an idea to a project level need not be an agonizing one. Many elements of the thought process can be isolated and captured in defining documents like the corporate (or project) vision statement. Those elements can then be utilized as simple and measurable metrics such a return on investment (ROI). Not only do the metrics help with the decision to go ahead, but they also can help with the allocation of resource decisions between multiple projects. The criteria can be both absolute and a relative measure of viability.

Opportunity Recognition, the First Gate

Since the overall goal of this text is to provide a framework (or model) of commercialization, I suggest that you consider a series of steps (or "gates"). The model dictates that each gate must be passed in order to move on to the next one.

The first gate is designed to ensure that the proper number of appropriate projects is distilled into "next steps." This requires two elements to be in place. The first is that the overall objectives or vision of the decision environment be articulated and understood. Without a cohesive vision or goal, the ability to make decisions consistent with that vision falls off quickly. Operationally, it is challenging to keep the clarity of the vision statement in front of the team that executes the projects.

The second is that a process is in place to allow an orderly yield of acceptable ideas to be sorted out. It is a bit tricky. An organization exists in a space defined by its vision, its people, technological competence, the resources available to it, and the markets in which it participates in commercial activity. Many times, these variables are used to define the need for new opportunities in terms of revenue, profits, or market share. This process must be timely enough to assimilate these attributes into the organization's capacity to absorb them.

One example is the pharmaceutical industry. Characterized by long development cycles and expensive research, it has become prone to utilizing in-licensed technology almost to a fault. Alternative strategies such as invention, joint ventures, and collaborations become secondary. A more balanced perspective might serve them well.

Distillation

Physical chemists have a tool that allows them to separate chemical mixtures by applying energy to different components that have relative volatility. It is called "distillation." The chemical process model serves as an analogy to understand how multiple ideas can be drawn down to usable projects.

There are many models for how this process of narrowing down the world of all ideas into usable projects might work. Those that allow quantification and offer clear resolution between alternatives are more effective than those that don't.

The model utilized in this text is based on a project I co-advised with Mac Banks at WPI and was led by one of my students, TJ Lynch. It was entitled "The Development and Application of Multiple Opportunity Analysis Tool for Entrepreneurs." (WPI- MQP, April, 2009).[1]

Lynch's motivation was a bit off the mark for what we are doing in that his original goal to develop the model was based on choosing the best of multiple, but disparate career opportunities facing him as a senior. He struggled to develop a method for sorting out which would be the best alternative for him to pursue. He quickly found that it was a complex task that consisted of multiple components and that the literature revealed very few tools with enough resolution to be useful.

His work utilized a weighted average methodology that is explained next, but is also visually shown in Figure 4-1.

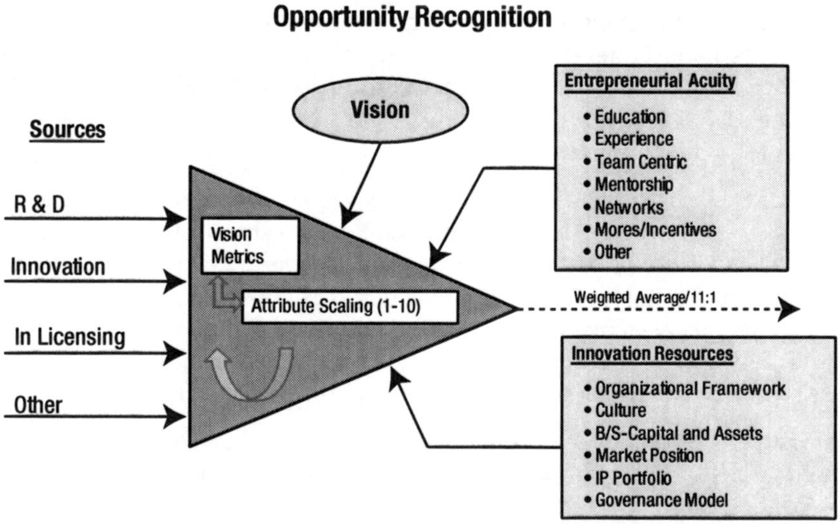

Figure 4-1. An opportunity recognition model

[1]https://www.wpi.edu/Pubs/E-project/Available/E-project-050509-091115/unrestricted/tjl-MQP-MOpA-Tool-D09.pdf

The model starts with numerical quantification of the vision or goals of the project shown in the center of Figure 4-1. It then attempts to "weight" the elements of the vision statement based on those goals. If, for example, the vision of the organization is to become the technological, low-cost market leader, higher attribution scores are allocated to those parameters. The dialogue between individuals or multiple groups party to the decisions that are used to assess the process of quantification can be rich and informative. This is especially true when alignment or agreement surfaces early. Opposing views take time and energy to resolve. A simple printed document standing alone rarely conveys its meaning. The dialogue is the meaningful part. In the context of the model, the attribution scores can be adjusted during the project to reflect changing conditions (such as the market) or changes in corporate strategy.

Once the project vision parameters are established, they are subjected to a matrix of functional measures such as technology, finance, operations, risk, and markets. This is labeled in Figure 4-1 as the Vision Metrics. These metrics become multipliers in the weighting calculation. The Lynch model utilizes 10 functional areas for analysis. Each subject or area becomes a basis for to a 10-question inquiry of strength and weakness within those areas. In technology, for example, they might query the number of patent disclosures per R&D employee. Measures like this become particularly relevant when the core business is technology dependent. Biotech or pharmaceuticals are good examples of this dependence. In retail-oriented projects, dollars spent to achieve market share through advertising or market incentives (discounts) might prevail.

The weighted average part of the model refers to the next process step. It is where the vision attributes are multiplied by the scores of the functional areas. The ensuing numerical score is then rank-ordered and weighted to reveal which projects should go forward.

Figure 4-2 shows the elements of the Lynch model. They include:

- A *functional element*. In this case it is Technology. It might be the functional areas of Markets, Competition, Finance, and so on.
- A *multiplier*, which is derived from the vision or goals statement for the project. In a technology-based effort, this number may be as high as 7 to 10.
- The project *rating score* (usually expressed in a series of relevant questions as a 1 through 10 rating). In this model, we try to discriminate one of three possibilities.
- The overall rating, which is a number that results from a multiplied effect of the multiplier times the score.

Opportunity Recognition (One element partial sample)

Functional area: Technology

(Multiplier = 9) derived from vision or goals statement

Sample Technology Questions (1-10)	Project 1	Project 2	Project 3
Number of Patents issued?	A1	A2	A3
Number of PhDs?	B1	B2	B3
Number of Successful Projects?	C1	C2	C3
Size of Facility?	D1	D2	D3
Budget?	E1	E2	E3
Percent of Revenue?	F1	F2	F3
Total rating score			
Multiplier (Weighing Factor) (1-10)	9	9	9
Overall Rating Score	225	400	150
Rank	2	1	3

Figure 4-2. Sample opportunity recognition model

When this process is applied throughout the functional areas and multiplied out, the importance that each area has to the overall project is identified and a weighted average figure of merit emerges. The various projects are then rank-ordered to clarify which projects are most aligned to the overall vision or goals.

Reading the literature about opportunity recognition, models that can discriminate one out of five candidate projects are considered good. The Lynch model demonstrates a resolution potential of one out of eleven, which is relatively amazing performance. The detailed workings of the model can be seen in the WPI MQP document.

As exciting as it may be to realize that a model can discriminate possible choices, the process is not without its limits. It relies on a quantification of goals or an articulation of the vision that can be numerically described. Does the use of multipliers have limits? Of course. One big drawback is this—with strict adherence to a set of multiplier rules dictated by the vision, you may miss unique "out of the box" disruptive or opportunistic leaps that intuition provides. It is a bit of a tight rope of decision making as to how much to rely on an analytical model and on intuition. At this point it seems that increased reliance on more disciplined and methodical approaches would yield better results.

Lego, the children's toy company, almost missed the market for robotics products because of internal uncertainty about their new innovation platform.[2] Workarounds like a statistical or "dashboard" approaches where a finite percentage of screened, intuitive opportunities are allowed in the

[2]*Innovation at the Lego Group*, IMD case study #382, Part B, 2008, Institute for Management Development, Lusanne, Switzerland.

mix by judgment or intuition alone might mitigate the risk of too rigid an approach. The point is that a "hunch" or "gut feel" can add significant value to the development of new ideas. Finding a mix of those decisions to the more orderly flow of ideas in a working model is the creative challenge. Without them, new and dynamic opportunities may be lost. Just look at the examples in the automotive industry. For years somewhat frivolous annual changes like tailfin additions and simple sheet metal shape changes superseded solid technological advances. No wonder the Prius came from outside of the Detroit industry mindset.

Multiple Funnels

Chapter 2 presented the argument for openness with regard to many sources of ideas. Then, models like the Lynch effort were presented as a tool to focus on distilling them into usable opportunities. Indeed, it is a disciplined winnowing down (or funnel) that is accomplished by applying that orderly approach. But, are the number of projects produced by these methods adequate? Is the capacity of the organization such that singular models are enough to produce sufficient opportunities? A unified model may not be sufficient. If it is, can it do so at an adequate rate? All this invites a series of interesting questions:

- Is the organization's track record of successful offerings of new products or services adequate? In simpler terms, what is the expected "hit rate"?
- Is the marketplace in which the organization operates dynamic enough to absorb new ideas? Is market share the predominate marker of success? Where are the competitive forces? What is their magnitude?
- Are external forces demanding change (that is, regulatory, environmental, corporate, and technological)?
- Is the organization and its resources properly positioned to achieve multiple entries in terms of people, capital, or technology?

Expectations

Beyond these questions, the risks of over- or under-estimating the number of "funnel" opportunities required for successful long-term value enhancements are significant. The numbers certainly help regulate the size and quantity of offerings required by the model. It may even open up a discussion that multiple models might be required to satisfy the operating needs of the parent organization.

Underestimating the requirements for new and innovative market offerings often means struggling against time. Expensive increases in overtime, FedEx overnight deliveries of parts and other operational costs—including people burnout to catch up—become part of the landscape. There is really no reason why a more orderly process cannot be employed. Certainly, it allows for loss of productivity and innovations. If acquisitions are utilized to achieve the goals, the time required for proper merging of the organizations becomes compromised.

If they are overestimated, issues such as introducing products or services before the financial lifecycle benefits of existing offerings are realized means that the full economic investment recoveries of the cycle are not realized. Cannibalizing existing products by accelerated obsolescence is sometimes considered dramatic and even effective in the marketplace, but the long-term value of such practices is questionable. An easy example of this is the constant release of new Microsoft operating systems. Although this strategy contributed to rapid rise of revenues (and profits) for the company, it also encouraged the development of "open" systems such as Linux as users became frustrated with the many changes offered by Microsoft.

With the sins of overestimating and understating the rate of new innovation developments in front of us, we have to wonder whether there is a "sweet spot" of resource allocation and commercialization rate that yields long- term value and results. Of course there is!

What are their markers? Companies that provide clues include 3M, with a culture that encourages internal innovation in its organization over a period of 100 years, and Emerson Electric, with its strict quarterly discipline and a long-term growth of 15% over 25 years. It is the focus of their new product and service offerings that are used in the creation and sustaining of long-term value propositions that produce the required results. Clearly, long- term proposition articulations and careful allocation of resources that are tuned to the longer-term visions become clear incentives for investment and market strength.

In the text entitled "Michael Porter: The Essential Guide to Competition and Strategy" (Harvard Business School Publishing, 2012), the author Joan Margetta contributes a Chapter 7 to "Continuity as the Enabler." She acknowledges Porter's five-step test of good strategy and process, but focuses on continuity as a test of strategy effectiveness. She likens it to a cooking metaphor that differentiates good strategy and process as a stew, not a stir fry. In the stew, there is time for the ingredients' flavors and essences to become integrated into the formulation. Analogies aside, the author then develops the rationale for long-term strategy and process. They include:

- Continuity reinforces a company's identity and allows it to build brands, reputation, and customer relationships.
- Continuity helps develop strong supply chains where suppliers and outside parties have time to develop optimum channels.
- Continuity increases the odds that employees throughout the organization will understand and contribute to the overall strategy.

Certainly, continuity and long-term value enhancement does not mean that an organization and its processes must stand still. Nimble and adaptive innovation must prevail to allow innovative competition. It does allow for stability of core values to be established and then that they be a platform for innovation to flourish. It also does not necessarily contribute to predicting future trends, but can allow inherent structural capacity to deal with changes in markets, technology, and customer preferences.

Chapters 1 and 2 focused on opening discovery processes so that there would be a robust and nimble sweep of possible ideas. It preached the sin of myopia toward any one source of innovative and new ideas. This chapter looked at the process of winnowing down those ideas to a useful flow of sources that satisfies the organization's capacity and need for them. In the Lynch model, this distillation process was shown to be able to discriminate ideas to an orderly process anchored to the core vision or ideas of the project and the organizational context in which they occur. Managing the flow of multiple funnels of opportunity to the dynamics of the organizational model's resource capacity becomes the new challenge.

Finally, we stepped back and looked at the overall metrics that control the process. Long-term value propositions and the continuity of strategies presented by Porter become the controlling elements of rate.

So, now you have a properly aligned set of ideas and are ready to move forward to another critical aspect of testing whether the ideas or projects have the probability to succeed. This next step is called "feasibility analysis" and might be considered the second gate.

CHAPTER

5

Feasibility Analysis

The Model

During the life of any project, there is a point where the logic of the project is challenged. Successfully passing this point (or gate) is a signal for going onto the next step in the process. This step is called feasibility analysis.

Although important as the feasibility analysis step is, it is fraught with uncertainty as the data gathered to make decisions is usually approximated and quite preliminary. Engineers in particular used to refer to this point as the "back of an envelope" check. By analogy, accountants will utilize a model in the form of a trial balance before committing the numbers to public scrutiny.

As apparent as this seems, the step is repeatedly ignored or done in a peripheral manner. Skipping this step can adversely affect the probability of success of the overall project. The risk of this exclusion by telling is shown in the story of the Smart Bottle project at Children's Hospital in Boston (CHB).

I was a consultant to the technology transfer office at CHB. One project they had underway before I arrived was a "bottle" that enabled feeding for certain neonates whose ability to breathe and suck fluids simultaneously (something "mature" systems can do) was compromised. Because of this condition, they were at risk of malnutrition or apnea. The current practice was to have a nurse cajole the child into doing both. The technology solution was a "smart bottle" that was a complex solution of pumps, valves, and computers. It worked, but its cost was about $5,000 per bottle and took over a million dollars to develop. The floor nurse who did the manual feeding had an incremental cost of $100. Had they done a feasibility check, they might have seen that the final product was 50 times more expensive than the current solution and likely more than many customers could afford.

Classical Perspective

In most feasibility analyses, there is a broad set of factors embraced in the term TELOS. The components are Technical, Economic, Legal, Operational, and Scheduling (capacity). I believe the categories fall into three areas. They include:

- Are they ready? Are there customers with identified needs and do we have the product (or services) and features at a cost that will attract their purchase?
- Are we ready? Do we have the secured technology, product, people, and information to supply those customer needs (and wants)?
- What happens if neither of those two conditions are met?

In preparation for this writing of this chapter, the model was exercised in several settings. In addition to providing a go/no-go option, it revealed the need for a third set of alternatives, entitled Hold and Abandon. More on that follows.

Figure 5-1 shows the model for feasibility analysis.

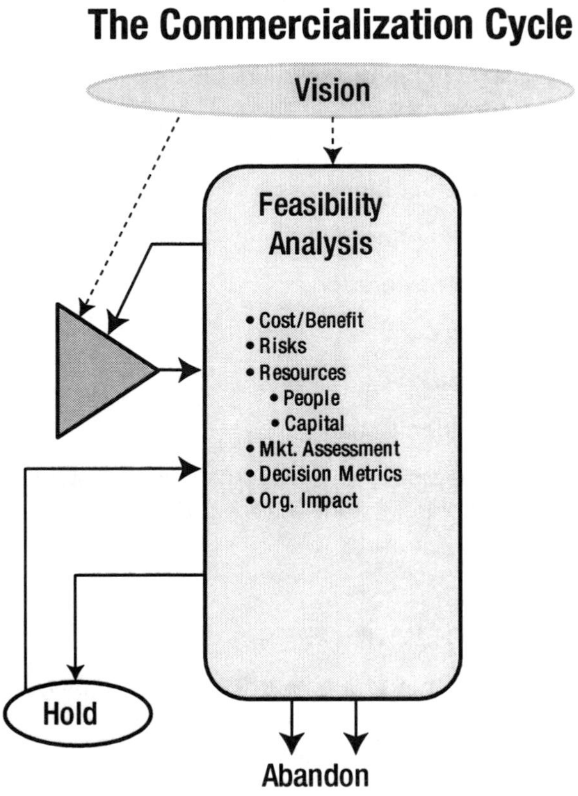

Figure 5-1. The commercialization cycle, feasibility analysis

---

Note The precedent for this discipline of doing a feasibility analysis is what used to be called the "back of the envelope" check. As a concept, it is a first order assessment of the functional areas of commercial decision making. In its implementation, it is a bit more challenging as it requires decisions to be made underusing uncertain data. This chapter provides a methodology for dealing with this uncertainty. As important as this step is, it is often dealt with in a cursory manner. As you'll see in some of the case studies, that's not a good idea. Case studies of the implication of that shortcut will be presented.

The Elements

There are six broad categories of the elements embraced that you must take into consideration during feasibility analysis. They can be broadly categorized as follows:

- Economic considerations
- Technical considerations
- Operational considerations
- Risks considerations
- Legal considerations
- Strategic considerations

When should you do a feasibility analysis? Although there is no specific agreement, it is clear that is should be done before major funds and other resources are committed.

Economic Considerations

Economic considerations rise to a high priority in this analysis. The category reveals the underlying cost models that are based on Bill of Materials roll ups in the case of tangible products and, in the case of services, Work Flow Models. They are certainly estimates but even as first order checks they help identify the feasibility of going ahead.

Materials considerations are usually bracketed by low and higher quantities. In established organizations, models of Standard Costs help normalize the accuracy of the estimates. Labor components also benefit from Standard Cost elements. In this modern world, the traditional make/buy considerations are highlighted as many products are outsourced globally.

There are also multiple external economic considerations such as the cost of money, Return on Investment (ROI) calculations, and of course the availability of capital as defined on the balance sheet or through the external capacity to borrow or raise capital through equity offerings. ROI is a simple calculation of gains realized by an investment. The formula is:

$$ROI = (Gains - Investment\ Costs) / Investment\ Costs$$

Beyond showing what gains are made from a particular investment, ROI allows you to compare possible returns between investments.

In combination, these considerations gate the decisions to go forward with the project. You need to consider capital expenditures, referred to as CAPEX, early as they usually have the additional component of requiring long lead items. An example is the need to acquire plant capacity and large manufacturing (or handling) equipment.

Money alone is not the final arbiter for testing feasibility. Market considerations, technical obsolescence, and aging plants and equipment are also combined to help rationalize the commitment to finance a project.

Technical Considerations

Of all issues confronting the progress of new projects, the dynamics of technological contributions seem most intriguing. Today, the rate of change of technological adoption is faster than any historical benchmarks before. Product lifecycles that were measured in yearly time constants are now presented in metrics measured in months. Enabling tools and processes are now so readily available. A classic example is rapid prototyping of parts. Historically, creating a physical prototype required months of drawings, and elaborate hand-crafted machining of metals and plastics. Today, a rapid prototyping machine directly converts a 3D AutoCAD (or SolidWorks) level rendering into parts in minutes. Even more significant is that the machines that accomplish this have seen an order of magnitude reduction in cost. Machines that cost $100,000 are now offered at levels of $1,000 and can sit on a desktop. This means that it is possible to create a working model of an idea or product in almost real time. Rather than wait months for a customer's reaction, a working model can be presented to them almost at inception. Even the tools that sketch out the part cans render encoded 3D renderings in almost real time (under the trade names of AutoCAD, SolidWorks, and so on)

An analogous ability to rapidly present ideas also exists in the electrical and chemical (materials) domains. Electrical circuits and chemical processes can be simulated on computers and results can be "tried" before expensive (and time-consuming) capital expenditure is made. A new class of "designer" materials can be formulated with specific functions. If all this weren't daunting enough, the global competitive environment has offered manufacturers the capacity to produce parts quickly. Turnarounds of days for parts from China are commonplace. The rate of change presents new challenges to markets and product development in that the changes can occur before the economic benefits of the product lifecycle occur.

Concurrent to advances in technology come vulnerabilities to the issues of infringement of intellectual property protection, as safeguarding of ideas has emerged. In their inception, patent laws were established to protect the fragility of startups as a platform for realizing the value of ideas for a period of 20 years. They provided a monopoly that protected such companies for the inventions. This model has been significantly challenged by emerging powers like China and India, where the business cultures disregard these laws.

This again seems to be changing as those powerful countries emerge into the world stage. They are now trying to form their own basis for IP protection and are launching national initiatives to catch up.

Operational Considerations

The operational readiness of an organization to embark on new projects centers on the current structure's capacity to absorb the requirements of the new ventures. They fall into two broad categories— plant/equipment and human resources. In common they are both "long-lead" items and require months, if not years, for a company to acquire and absorb. Operational issues include the capacity of the physical plant, equipment, people, and financial balance sheets to support the proposed project(s).

An example of operational feasibility might involve having enough computer capacity to support more customers, additional online features, or simply maintain an appropriate speed of response to operational needs.

As early as the feasibility analysis is, it is common to explore future issues in a time-based format. A common tool is the Gantt chart (Figure 5-2). It was invented in 1910 by Henry Gantt. It sets out tasks to be accomplished in an orderly and time-scheduled manner. It also identifies interdependencies of decisions and places points where feedback is required. In addition to being a scheduling device, it also allows you to track expenses and salaries required during a project. The personal computer added significantly to the usefulness of the tool as it allowed more complex models to be processed more readily. Microsoft has offered a program called Project in its Office Suite that simplifies the development of Gantt charts. There are also alternative tools available to complete these tasks.

Although the project management usefulness of these charts in project management is unquestioned, it performs an important set of tasks for commercial success by requiring the identification and quantification of long- lead items. These generally fall into the category of capital expenditures (CAPEX), which have magnitude and exhibit long-time constants for delivery. Examples include purchasing buildings, hiring people, and buying machine tools. It is important to the overall success of the project that commitments are made for design and implementation even before the project is completely authorized.

Risk: Considerations

Managing risks is an essential element of any modern commercial enterprise. Delineating risks (and offering mitigation alternatives) are markers for a well-presented business plan. At minimum, they suggest that management of the project risks have been thought through. It is also understood that mitigation alternatives that are probabilistic in nature might not work.

Note The largest risk that an enterprise can confront is not addressing the possibility that unforeseen, undesirable events occur. The ability to frame and resolve risks is the backbone of good management practice.

In certain financial domains, for example, there is a family of reporting risks. Classic is the Securities and Exchange Commission (SEC), which requires a chapter of any offering prospectus to identify primary risks. In industries where safety is paramount, there is also a requirement of delineating applied risk factors. Accounting firms in their auditing practices identify risk factors in qualifying a business's ability to be an "ongoing entity." Medical and aviation are easily identified as areas where the delineated risk issues are important.

Some risks are mitigated by third parties. An example is casualty loss due to fire or natural events (such as floods). Insurance companies that base risk coverage on actuarial models deal with risks as a monetary transaction based on assessed premiums. Other risk categories, such as creditworthiness of customer contracts, are managed by "good' business practices. Dun & Bradstreet, the financial services company, offers courses in determining creditworthiness of potential creditors. Property mortgage issuers are normally looking at credit histories of the potential mortgagees, and they also utilize technical inspections of the property to their assessment process.

Legal Risk

There are specific risks that fall into this category. Let's look at several examples.

Contractual. Most commercial transactions are bound by common law contractual instruments. Sales, for example, are normally are captured in standard legal instruments but are bound by the rules of the Uniform Commercial Code. These terms are sometimes printed directly on the back of the purchase order form or transaction slip. Risks are carefully bound but are challenged by non-delivery or other breach terms.

Also in this category are internal contracts such as employment and non-compete issues. Although routine in nature, certain nuances like severance clauses of employment contracts are areas of careful negotiation. At higher levels of employment, compensation committees of the board of directors set guidelines for these discussions.

Intellectual Property (IP). This area of legal risk is also one of strategic importance. Not only is the crafting of patent and copyright instruments the domain of legal discipline, but the maintenance and defense against the infringement of the IP becomes a key element of external risk of the project. Sometimes the issues in this category challenge the branding image of the product protected by copyright.

Governance. Boards of directors in modern companies have clear fiduciary responsibilities to the shareholders and owners of the company. With the advent of the Sarbanes-Oxley Act and its attention to the roles of the individual board members, it became one more area of litigious activity. Prior to the law, board members could hide behind "not knowing" the issues. Classic was the Enron corporation case where illegal off-balance sheet transactions were used to hide losses and non-performing assets. This defense of "not knowing" is no longer tenable. Broad issues of corporate ethics and integrity also flow through the governance model of the company. Director insurances and internal by law indemnification barriers tend to mitigate the risks, but shareholder suits are still common.

Financial: Risk

Of all areas of risk, managing financial risk rises to the top of any list. Although projects look at several possible channels for this vulnerability, weakness in the balance sheet that results in cash deficiencies tend to trigger maximum exposure to failure. It is likened to an aircraft that exhibits fuel exhaustion. The negative glide slope of descent without the engine's thrust creates a predictable, although not always successful, landing. Some of the more prevalent forms of financial risk include:

Cash Planning. This can be accomplished with sufficient accuracy in the first years of a project. Allowances for unplanned fluctuations can be mitigated by creating cash reserves early in the company or project history. Acquiring capital infusions from traditional sources such as angel or venture capital firms is a good alternative but can be delayed due the internal decision processes of the source. Exceeding the covenants of bank loans can limit the availability of cash. Slow payments or weak credit from receivables sources can negatively impact the flow of cash in a company. Careless or even unplanned expenditures such as overtime or FedEx expenses (instead of less costly alternatives) can all contribute to insufficient cash.

It's argued that financial disciplines should be applied early in a project or a company's existence. Clearly, early new organizations can't easily prioritize the need for a formal accounting or financial system. An adequate alternative is to engage a part-time Chief Financial Officer (CFO) to support the accounting functions but also to participate in external financial decisions such as negotiating investment capital infusions and CAPEX deliberations.

Commercial and Market Risk

The core of commercial risk evolves from the multiple elements that affect the sales transaction. Clearly, marketing influences such as branding, packaging, design, pricing, and distribution methods enter into this. Each contains an element of risk. Competitive pressures impact the actual purchasing decision. It is an oversimplification to just refer to these elements as competition. It is more complex than that.

In the 1990s, the Miller Heiman company released a concept called Strategic Selling. Central to the concept was the identification of the multiple purchasing influences flowing through a central spokesperson such as a buyer or a procurement specialist, called the Technical Buyer, the Economic Buyer, and many others. Each "type" has a differing set of objectives in the purchasing discussion. Utilizing a "blue sheet" format, each influence was identified and then a specific strategy for confronting them was formulated. Many times individual influences were in conflict and thus needed the selling sources to both delineate and manage them. Through processes like this, the rate of successful closures increases. Of importance is the fact that any buying influence can negate the closure of the sale. Managing that process is the domain of the sales management function.

Beyond examples like the Miller Heiman, which can be implemented through proper organizational resource allocation and training, there exists a range of risks that are beyond the corporate models. Examples include new breakthrough technologies, emergence of new competitors, regulatory mandate changes, and so on. Certainly it is within the scope of management to be aware of the possibilities that can occur (and perhaps offer contingency plans to offset these risks).

Regulatory Risks

Beyond rules imposed on basic industries, the impact of regulatory change becomes even more significant. Five- to ten-year FDA approvals cycles for pharmaceuticals and the financial impact of these processes are almost mind numbing and certainly stand squarely in the face of rapid change and innovation. At a minimum, that leaves significant and disruptive change to the domain of large companies.

Regulatory issues pervade every aspect of modern business and you must pay attention to them. Whether it is OSHA governing workplace safety or Sarbanes-Oxley setting standards for board governance issues, each contains uncertainty of change, and thus risk. Once again it becomes the domain of management to sort that out and to establish the priority of those that will impact future performance. It is interesting to note that when the Sarbanes-Oxley Act was cast, it actually provided for the formality of risk assessment and mitigation to the board.

Many regulatory aspects are complex and seem to be in a constant flux of change. There are a group of specialists in each area who serve as consultants to industry. They offer up-to-date information, as well as broad solutions gleaned from multiple clients.

We live in a world of global competition. Regulatory issues are not consistent and may even be in conflict. An example is the automotive industry where high-end cars like Mercedes and BMW couldn't be imported into the United States because they didn't meet basic safety regulations like headlight, bumper, and windshields specifications. This discontinuity also invites opportunity. The U.S. Food and Drug Administration is the most stringent drug regulatory agency in the world. Some companies thus develop early market drugs in other places in the world. The United States is the largest market for most products, so eventually these differences tend to normalize about its standards.

Risks: A Summary

Risk factors are inherent in all aspects of commerce and commercialization. Offsetting these risks at minimum entails weighing the rewards or the larger upside potential. This potential is usually measured by the return on equity (ROE) of funds invested in the equity of the organization. Multiples of 10 are usually the primary thresholds to professional investors like venture capital fund managers.

Agreeing that because a landscape of risk categories confronts any ongoing businesses, it becomes incumbent on the managing forces of company managers to identify those that are relevant and present a plan of mitigation of those risks to potential investors, employees, and customers as a condition of their interest, involvement, and participation.

Additional Elements: Outcomes of the Feasibility Analysis

Successful feasibility analysis occurs early in a project. Its purpose is to flag appropriate projects to invest additional resources on the path to commercialization. It is a crisp decision point made under the uncertainty of less than complete data. It accomplishes additional benefits, discussed next.

Rejection

Feasibility analysis allows you to terminate a project cleanly in comparison to the stated vision or project's purpose. This allows you to conserve funds and avoid market mistakes. Most important is that you can do this in an orderly and disciplined manner. Feasibility analysis helps you to avoid or at least diminish the use of subjective and non-factual information. As straightforward as this step is, it still becomes a subjective test of management strength to implement.

Hold

In a perfect world, projects would either be a go or no-go result of the feasibility analysis effort. In reality there is a significant amount of ambiguity, some of which can be quantified. For example, a project may contain an electronic circuit comprised of multiple components. In order for the overall product to be considered viable, it requires that the circuit board meet certain cost objectives. In this case, certain components are simply too expensive. It is known that the cost of them is anticipated to drop. It is therefore quite appropriate to put the project on "hold" until that cost objective can be achieved.

All this carries a responsibility, which is to capture the legacy of information that has been generated in the process. Projects that make it to "hold" have met a variety of gates (and expenses) to get to that place. "Hold" is not a rejection step, but rather contains a catalog of projects that are waiting for a specific set of outside conditions to change before they can move forward. The cataloging process has to be more than anecdotal and calls for a librarian's discipline of catalogs and flags.

An example is a project that contains electronic components that are new or evolving. This makes them expensive and may move a feasible project to one that is not financially feasible. In this example, the price of the electronic components must be monitored on a periodic basis to observe when they drop to a point it would allow the project to be reconsidered. Putting the project into a "hold" position with a marker to track component prices would be an excellent use of the "hold" position.

In some cases, "hold" allows broad outside conditions such as regulatory, market, or even modifications of the corporate vision to change in a manner that sets up a more favorable environment for a project to move forward. Over time, it becomes statistical in that it's affected by the amount of projects a company allows and thus has the capacity to support. The yield of projects leaving the "hold" area and, combined with the yield of the Opportunity Recognition phase, measures how robust the pipeline for the new projects is. In the end, the corporate entity relies on a stream of new projects to be able to compete in the area of new products and services. Shorter product (and service) lifecycles and global competition make the management of the pipeline dynamics more important than it has been historically.

Summary

Feasibility analysis is not a new discipline. The early craftsmen, marketers, and engineers used primitive indicators such as notes on the back of envelope and other limited tools to help decide whether to move a project ahead. Probably intuition played an important role. Today, the stakes are much higher. The rate of change in technology, the element of global competition and, of course, the Internet leaves less room for error and demands a faster response. The conventional TELOS approach seems to have added a dimension of orderly response to it, but is still falling short of today's requirements.

In this chapter I proposed that a methodology that not only looks at a) how ready we are and b) how ready and available the markets (and customers) are but also c) provides for an orderly rejection of projects and d) controls the suspension ("hold") of projects is required. The penalty of not confronting these issues results in loss of markets (and following revenue for growth), and has the potential to squander precious resources of time, people, capital, and company reputation.

Once you have decided to move the project forward, questions how to strategically move forward to commercial reality become the next gate or threshold to overcome.

CHAPTER 6

The Project Plan

Decisions, the Precursor

Do all organizations plan? How formal are the processes? Are they effective? These and many other questions sets the stage for us to look at the process of planning. This chapter looks at how the operative steps can be used to improve the probability of success of a commercialization project.

To appreciate the context for this, consider again airplane pilots and the sophistication of the planes they fly. The early barnstormers didn't plan their flights very methodically. Their human-machine interface was perfect in that there was a direct coupling between the pilot and aircraft controls. Perhaps a thumb to the wind might have been in order. On the other hand, the chief pilot of a Boeing 787 is utilizing sophisticated computer-driven data complete with weather information from around the world. The earlier airplane can only go 100 miles (if you're lucky); the other is capable of circumnavigating the globe. Which model fits best?

Why Planning?

The simple answer to "why plan?" is that it can improve the probability of success of a project and promises to reduce losses and inefficiency. In the modern world of increased global competition and accelerating rate of change, improving the odds of success is the desired goal.

Planning exists in the space between project uncertainty and improved positive outcomes. The degree of formality, detail, and depth are also determined by the maturity of the organizational context in which the project exists. Whatever the level of project planning sophistication employed, an overriding benefit of the process is the internal communication of goals, organizational interdependence, and measurable outcomes that are positive attributes of the process. Let's look at the details to better understand this.

The Details

A well-thought-out feasibility analysis offers direction as to how best to proceed to a viable commercialization outcome. It also reveals a common weakness in how some folks approach this. It is that there is an organization presumption that there is a favored pathway to market. In many organizations, for example, it is common to exploit new ideas with starting a new company, yet this may be the most unlikely path in terms of risk and market penetration capacity.

Multiple alternatives should be constantly examined to see which path best serves the particular project. A partial list is shown in Table 6-1.

Table 6-1. Alternative Commercialization Pathways

Alternative	Attributes
Startup	Allows separation from a parent organization and its brand. Offers the flexibility of a specific organizational model focused on the project. Risks include the uncertainty of new team and capitalization model.
Licensing	Offers a quick path for a new technology. Carries the financial overhead of a license fee, while relieving the cost of R&D and regulatory acceptance.
Joint Ventures	In many ways an attractive alternative in that it draws from multiple organizations. It loses the strength of the internal learning curves and poses the risk of mixing two diverse cultures.
Franchise	Offers an alternative project funding through fees and royalties. Carries the responsibility of providing quality and new products to sustain the franchise value.
M&A	Selling the project to another entity. Brings immediate liquidity but compromises the market preens of the new product.

Although this is a partial list, it reveals that each pathway has different dynamics and responsibilities let alone outcomes. Some of the decisions are clearly industry specific. An example is the pharmaceutical industry, which has long favored the duality of internal R&D and licensing. How do we decide?

Deciding

There are certain broad metrics that control decision making at this stage of the process. As the journey toward commercialization evolves and additional information becomes available, it may offer a point to review and even retrace the initial steps. The positive aspect of this type of planning is that few expenditures of human and/or financial capital have been invested. Some of the early considerations include:

- Corporate vision. Clearly the culture and goals of the sponsoring organization achieve first consideration. There are exceptions, of course, as with the rogue group of 17 engineers and staff from IBM who stood in the face of the then corporate identity of only making large mainframe computers to develop the PC as we know it today. They rented a garage in Boca Raton, FL, procured parts from a local electronics parts store believed to be Radio Shack, and began their work. This anecdotal story is offset by many "counterculture" projects that failed.
- Industry mores. The folks who work in the fast food and retail industries operate in a different culture than those working in the computer mainframe or scientific hardware businesses. Can that defining constraint be ignored? Certainly, crossover products and market disrupters occur frequently. When viewed on the basis of final probability of success, one ponders at how large a cultural price can be absorbed.
- Human capital. Possibly the most critical factor and most commonly ignored. Careful assessment of the people required to successfully achieve project goals needs to be done early. It takes time to find and assimilate the skills and experience required to begin and then prevail in the market. There are professional outside firms that can be hired if internal human resource functions aren't available.
- Financial resources. Both careful analysis of the requirements and a review of the alternatives is required. In larger, public-traded entities there are alternatives for funding that come from professional resources such as public offerings of liquidity that takes time to acquire.
- CAPEX. Long-lead items such as the acquisition capital equipment and space take time and require careful scrutiny. It is tightly couples to the acquisition of capital. The "make-buy" decisions that surround this have now led to sourcing from China and the other countries of the Far East.

Drilling Down

As the detail of the planning process unfolds, a series of "information layers" is created. They literally become the chapters of the formal document. They include:

- Project objectives. In the multiple steps of the commercialization cycle, there is an ongoing connection to the overall vision of the company.
- In the decisions about opportunity recognition, for example, the vision statement is actually quantified and used to weight the functional evaluations. The result is a "weighted average" response that yields high resolution.
- Assuming that the corporate vision becomes the basis for measuring the overall organizational performance, aligning a given project to those goals enhances the overall performance metrics. Given that, there are two caveats:
  - Possibly the greatest benefit of the linkage to the corporate vision is the level of communication required to ensure a match to the goals of the central vision process. This requires significant dialogue and connecting the project objectives to the central plan. There are many benefits, including the alignment of efforts needed to ensure that adequate resources are allocated to the project. It allows the members of project team to engage at the highest level of the organization.
  - All this is not without issues. Strict adherence to the central corporate goals limits the energy of disruptive and bold (innovative?) change. Some percentage of new products and new market creation require this. Nowhere is this more apparent than the expensive world of technical innovation, where the rate of change is constantly increasing. Sometimes a "dashboard" approach that monitors an agreed amount of measurable disruption is utilized to show allowed deviations from the central themes.
- Concept definition. With the strategic issues aligned and articulated, an important next step begins. It is where the actual boundaries are defined. Some of the areas include the definition of the team, initial cost constraints, budgets,

cost/benefit arguments, make/buy decisions, schedules, deliverables, risk factors, and success metrics. Tools such as written software specifications, detailed bills of materials, and process flow charts become invaluable.

- Approach. There are several methods of planning that are broadly characterized by the degree of formality and timing they utilize. Certain aspects such as financial modeling follow those groupings. In a low risk, somewhat predicable environment, Net Present Value (NPV) is as good a financial modeling tool as any. NPV is a calculation whereby the value of money in a future time period is brought back to a current time.
- Martha Amram, in her book entitled *Value Sweep: Mapping Growth Across Assets*, argues that more complex and fast-changing projects require a decision-tree format that compares alternative approaches. The "tree" model allows comparisons that are more useful for upper management decision making, as it gives management more timely comparative information.
- Methods that are in use include:
  - Traditional project planning. Uses linear multistep planning for the overall project.
  - Incremental planning. As its name implies, it breaks the project into phases and delivers them sequentially. This reduces the risk of delivering everything at once and becoming inaccurate as the data and assumptions change over time.
  - Iterative. Allows the project to absorb information in a timely manner and then have the project evolve. Susceptible to loss of focus and urgency as outside conditions or internal teams change.
  - Adaptive. Also allows the project to change over time and as information quality increases. In this case, it is planned with multiple phases of effort in advance.
  - Extreme. As implied, this category allows constant scrutiny by various stakeholders, such as customers and management. It's susceptible to loss of direction and requires close team coordination.

Over and Under

The previous section points out that there are many levels of planning and tools that may be employed. They vary in detail, scope, and frequency of review. In common, they focus on a set of deliverables that map out the direction of a given project.

The style of planning varies most with the stage of growth of the parent organization. Clearly a startup or early stage company will employ a more concise, less elaborate format. A more mature organization will necessarily allow a more detailed, more thoroughly reviewed process. Each has its positive and negative attributes. What they must have in common is the ability to guide the organization in an orderly manner that allows it to prevail in its chosen commercial space. The outcomes can certainly be measured and the choice of process can thus be modified to achieve best results as measured in market success.

Whatever process is employed, they are susceptible to common foibles. The most common is when the planning process takes on a life of its own and becomes a bureaucratic function that only allows marginal results in terms of commercial success. There are several insidious issues that threaten the success of a given process. They include:

- Scope creep. Changes in project scope happen all the time. When the project becomes more complicated than needed, this form of creep threatens its success and possibly interferes with its outcomes.
- Hope creep. This is where the ultimate human behavior of hiding or distortion enters the dialogue. Inevitable and quantifiable measurement of progress becomes important to counteract this. Good project management becomes critical.
- Effort creep. When the task becomes more complicated than anticipated. Individuals tend to invest more hours but measurable goals fall behind.

In a thoughtful article published at the Kelley School of Business (Elsevier 2013 titled "Lies, Dammed Lies and Project Plans: Recurring Human Errors That Can Ruin the Project Planning Process," Jeffery Pinto enlarges this idea by citing that "Project-based work has become a critical component of global industrial activity. Unfortunately, the track record for project development has not been strong." He further defines failure as discontinuation of the project or even cancellation and cited cost and time overruns. More often than not the project teams are tasked with trying to salvage value and profitability. Somewhat of a grim scenario!

He further lists the "seven deadly sins" of human behavior that contribute to those negative outcomes. They range from external changes (such as a sales team forecast promoting the product early) to unrealistic timelines, poor change control to adapt the project underway, poor project management training, and so on. The article concludes the need for better skills training, curtailing systemic errors, and broader allowance for rework cycles as future directions.

In an article titled "How to Fail in Project Management (Without Really Trying)" Pinto and co-author Om Kharbndo elaborate on the failure modes of project management and challenge the readers to be willing to learn from their own mistakes.[1] The list is extensive and includes:

- Ignore the external environment and stakeholders
- Push new technology too quickly
- Don't bother building fallback plans
- When problems occur, shoot the most visible
- Let ideas starve to death from inertia
- Don't bother to conduct feasibility studies
- Never admit a project is a failure
- Never conduct post-failure reviews
- Never exercise tradeoffs
- Make sure a project is run by weak leaders

He concludes by observing that past failure (or success) is not a predictor future outcome. There are lessons to be learned, but only if we are willing to find and examine the problems.

The Tools

Beyond the basic measurement tools of financial performance, such as Return on Investment (ROI), which measures the overall ratio of investment and returns, there are other measures. The Net Present Value normalizes items to return over time so that multiple projects can be compared independent of the time to realize their benefits. The Internal Rate of Return (IRR) is useful to quantify the rate by which cash resources flow through an organization. This last tool becomes important when you're considering large capital expenditures such as plants and equipment.

[1] *Business Horizons*, July–August 1996.

There are a litany of tools such as milestone charts, market share analysis, and ratios of headcount to costs that become important parameters. Perhaps the most significant of all these tools derives from the field of project management. Three significant ones include:

- The Gantt chart. Formulated by Henry Gantt in 1910, it is still commonly used today. Companies like Microsoft offer software products like Project to derive and control this method. The Gantt chart is a listing of tasks, the individual(s) who will implement them, and the budget implications of these assignments. It allows you to visually monitor the progress of the deliverables. It is not uncommon to see these charts enlarged and posted on project centers as their graphical features are useful.
- The Critical Path Method (CPM). Created by DuPont and The Remington Rand Corporation in the 1950s as a tool to monitor and control maintenance tasks. It allows the actual time to perform the tasks to be determined by the time reported to complete them.
- Program Evaluation and Review Technique (PERT). Also developed in the 1950s by the Booz Allen Company and the U.S. Navy to monitor and control complex defense projects. It graphically portrayed time-based tasks and their interdependence to related tasks.

There were many attempts to consolidate these tools and present a common format. In 2006, the American Association of Cost Engineers released the Total Cost Management Framework. It has not been universally adapted. As the world continues to move to a global context, there became international standards offered for use. The International Project Management Association (IPMA), for example, defined the IPMA Competence Baseline (ICB) tool to help standardize the terms and context utilized in international project planning.

When I queried Google with the words "project planning," 109 million entries were reported. Cleary, this is an indirect measure of the size of the task. It amplifies the importance of adapting the planning process to the needs of a given organization.

To Plan or Not

Perhaps it's a bit dramatic to paraphrase the line from Shakespeare's *Hamlet* where the lead character (Hamlet) contemplates the value of life and potential suicide. Planning reaches beyond the polarity of whether or not to perform a planning discipline as part of a commercialization effort to define which process serves the decision process best. It is clear that insufficient planning effort exposes the projects under consideration to risks of outright failure, cost overruns, unexpected delays, missed deliverables, and poor internal communication of goals among the participants in the process and missed market opportunities. It also provides a road map to the expectations of the deliverables and resources required. Critical among these are the long lead items that can bedevil projects by the light of time required to obtain them. You can't "push a string" in attempt to correct the time-based issues.

Further, whatever path a planned project embraces, it is clear that there is a need to match levels of detail, sophistication, and documentation to the needs of the enabling organization.

The penalty of squandering precious resources like money and talent are that windows of market opportunity may pass. There are fewer resources available to pursue more promising avenues. Softer issues like weakening the brand name, employee morale, and softer balance sheet financials can take years to overcome. The incentive is to do it right the first time.

Whatever the nature of the planning process that's identified and executed, the required functional elements that ensure the success of the project must be assembled in place. Perhaps the most complex of these is to look at the issues surrounding the marketing of the product or services. Let's look at that in the next chapter.

CHAPTER

7

To Market or Not...

The Context for Commercialization

Any business text will quickly identify the importance of marketing in a commercial project, and so will I. But here the issues are a bit more complex. It becomes exceedingly important to delineate not only the market opportunity but also the appropriate sales engines—including startups new divisions, franchising, licensing, joint ventures, and so on—as platforms to recognize its potential throughout the product lifecycle.

Marketing is a critical aspect of understanding (and acting upon) how a given product or service integrates into the customer's perceived needs. It is not a singular function but rather a range of interactions both inside and outside of the organization. Marketing's position may be best observed in Figure 7-1.

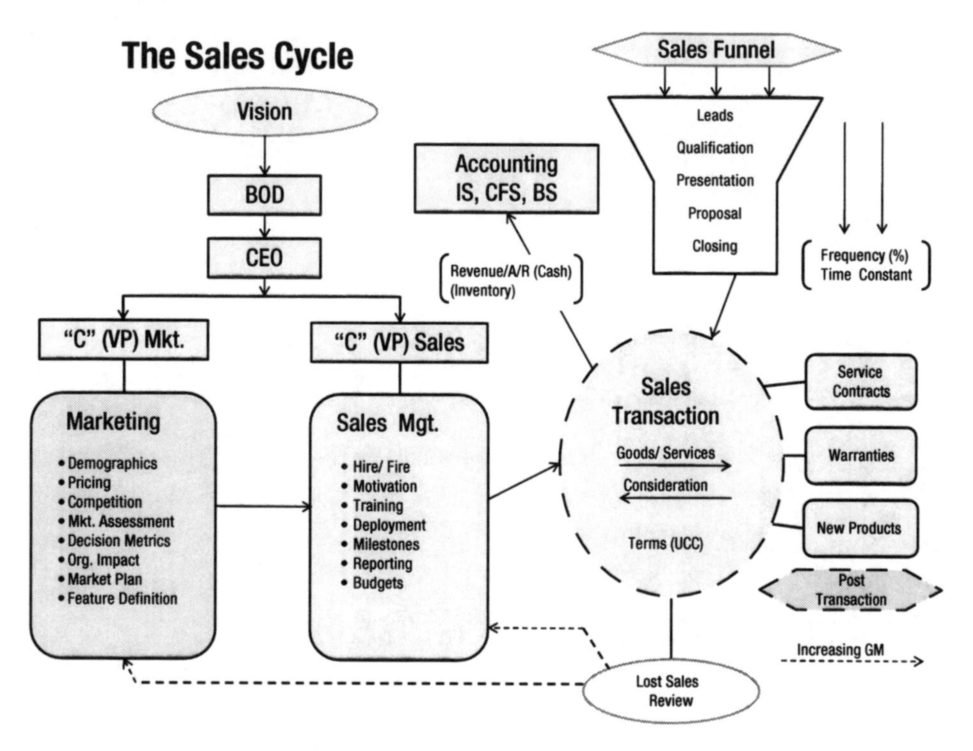

Figure 7-1. Sales/marketing interaction

In ocean sailboat racing, it is interesting to observe the role of the skipper, also known as the helmsman. Usually, it is the owner or a similar commanding figure exercising the helm, looking quite authoritative and sometimes shouting orders for sail changes or directional headings with great fanfare. It's really quite a photo op, and the implication is that the winning strategy is derived from that behavior.

The unsung hero in this drama is the tactician, who sits below the skipper and is constantly calculating wind direction, sea forces, and relative boat positions. The tacticians are typically hunched over a laptop or other computing device and succinctly issuing alternatives for the captain and crew to follow. The race is usually won by this function.

So it is with a corporation, where the CEO and CMO (Chief Marketing Officer) play analogous roles of the skipper and tactician on the boat. Winning has something to do with the directions set by marketing.

The End of Marketing?

In the period of 2005 to 2009, there were a series of devastating articles predicting the demise of traditional marketing. Fredrick Webster, Jr. and others published an article in the 2005 *Sloan Management Review* entitled the "The Decline and Dispersion of Marketing Competence." In it, they reported corporate CMO tenures of less than two years. Marketing budgets fell to single digits and the percentage of revenue reported and staff turnover was rampant. In the *McKinsey Quarterly* of the same period, half of a group of European CEOs were "unimpressed by their CMO's performance and felt they 'lacked business acumen'." In a similar article published in the *Sloan Management Review* (Summer 2008), Yoram Wind argued that the discipline of marketing "hasn't kept up" with the rapid changes of the 21st Century. He cited Tom Freidman's concept of the flat world, the rise of China, the Internet, and social awareness as the markers of this observation.

What really happened is that the concept of commercial business as we knew it was undergoing a major disruptive set of changes. The concept of marketing had to adapt. In a 2014 Gartner report, called "Key Findings from U.S. Digital Marketing Spending," James Rivera and Rob van der Mullen found marketing budgets in 315 companies were predicted to rise 8% and that 14% of the respondents planned to spend over 15% of their revenue on marketing budgets and salaries. Laura McLellan, vice president of research at Gartner says, "The line between digital and traditional marketing continues to blur. For marketers in 2014 it is less about digital marketing than marketing in a digital world." This is a significant increase and bears further consideration.

How could this happen? There seems to be multiple contributing elements. They include:

- Breathtaking extension of the Internet into every walk of life.
- Mind-numbing extension of smart phones, tablets, and computer usage.
- The movement to a global competition for customer, resources, and markets.
- The rapid extension of social media such as Facebook and Twitter.
- Kindle, Nook, and MP3 in the publishing arenas.
- Success of Amazon and other e-commerce.
- Trained specialists in web design, viral marketing, and enabling technologies.
- New market search engines, such as Google.

Eric Von Hipple, the MIT professor of Innovation, wrote a text entitled *Democratizing Innovation,*[1] in which he observed that we are rapidly coming to the place where all of us can create our own commercial solutions. He used the example of how we can create concepts on our personal computers using AutoCAD or SolidWorks and then immediately transfer those ideas into parts through the use of rapid prototyping devices. These machines can be purchased on the open market for less than $1,000. Although the output of these devices is limited in scale, materials, and accuracy, it doesn't take too much imagination to see where this might lead.

But how does all this connect to the commercialization of ideas? It is clear that as we discern how to increase the probability of success, these new channels and marketing tools must be brought into focus. Relying on older marketing tools is no longer sufficient.

The New Marketing Model

Within the context of a new model of the marketing paradigm created by the changes outlined previously, there is a need to look at commercial opportunities in a different context. A starting point is to look at the model from its organizational function, as shown in Figure 7-2.

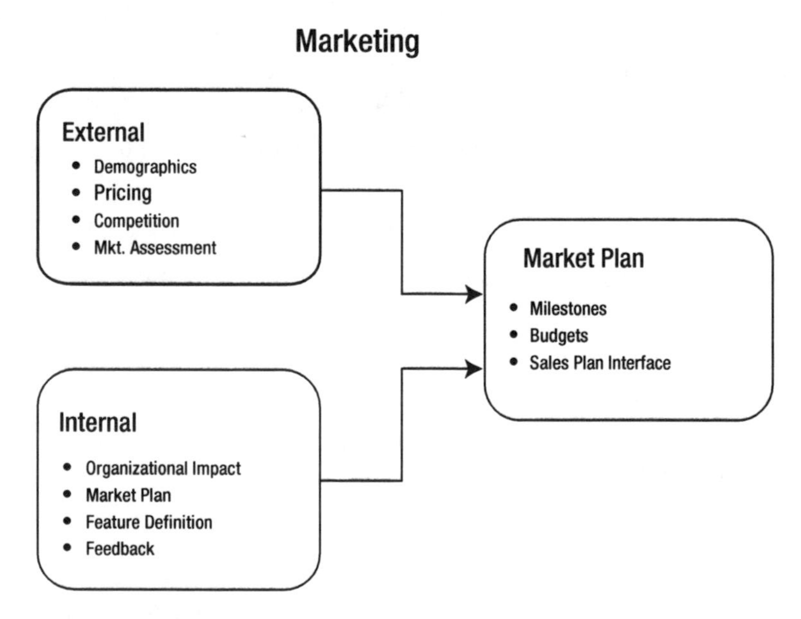

Figure 7-2. Marketing function

[1]*MIT Press*. Creative Commons License, 2005.

Of the many aspects of marketing that are of interest, two of them have particular focus. The first is a term similar to what pilots call situational awareness. It relates to a pilot's ability to relate to the environment around her. It measures how well individuals respond to variable changes in that space. Markets rely on that ability. A tool that is employed is a matrix of the Strengths, Weaknesses, Opportunities, and Threats (SWOT). It is graphically represented by models such as the one shown in Figure 7-3.

SWOT Analysis

Strengths (Internal) 1. 2. 3. (Helpful in Achieving Goals)	Weaknesses (Internal) 1. 2. 3. (Harmful in Achieving Goals)
Opportunities 1. 2. 3. (Helpful in Achieving Goals)	Threats 1. 2. 3. (Harmful in Achieving Goals)

Figure 7-3. SWOT analysis format

The SWOT analysis is a mainstay of market assessments. In reading many business plans, I often find them presented but with static (one-time) information. In reality, a more accurate perspective would be a kaleidoscope of snapshots along a project's journey to commercialization. Noting the dynamic changes (and how the project team reacted to them) becomes an important marker. It's a sensitive indicator of the situational awareness confronting the projects. Tracking changes along the way becomes a powerful perspective about the project's viability.

There is a potential that the marketing and sales groups operate in conflict. This derives from a lack of clarity about the two functions from upper management. It is the clear responsibility of upper management to define each function to offset these possible conflicts.

To clarify the potential issues between sales and marketing, it is useful to look at some of the critical components of each.

Demographics

In more traditional marketing models, demographics is the focus of what the first contribution the marketing function adds to the success. Demographics helps define:

- Who are the customers?
- Where do they live?
- What do they buy?
- How do they make purchases?
- What are they will to pay?

The list of defining attributes further includes age, gender, marital status, purchasing power, and so on. Included is the dynamic model of what changes or trends the project is undergoing. The information is creatively synthesized into a demographic model of a hypothetical aggregate customer. This information is then used as the basis of the marketing plan and formulation of a brand strategy. Like the SWOT analysis, this is not a one-time static vantage. It is a dynamic ever-changing landscape. The creative part is to utilizing this information by observing changes and trends and reacting to them.

Renee DiResta wrote an essay on O'Reilly's RADAR entitled "Demographics Are Dead: The New Technical Face of Marketing."[2] In the article, she claims that marketing has been transformed from a primarily creative process to an increasingly data-driven discipline with strong technological underpinnings. Although marketing still has a primary mission of affording a connection to the customer, the pathways to the buying decision and the resulting buying patterns have seen a fundamental change. DiResta argues that the "Old Marketing used a 'spray and pray' model aimed at a relatively passive customer base."

Advances in data mining have improved to the point where marketers can develop highly specific profiles of customers at the individual level using data drawn from actual costumer behaviors. Amazon is a good example. They maintain a living database of all transactions and inquiries made by the customers. When you log on to the system, Amazon suggests books and other products that are extensions of your personally indicated interests, as judged by your history of previous purchases.

Added to this new approach are the breathtaking deployments of laptops, tablets, and smart phones to consumers worldwide. They allow the consumers to become incredibly knowledgeable in product (and service) offerings as well as in pricing and supply chain availability. With this, traditional consumer loyalty has evaporated and a ferocious new competition occurs at the product/consumer interface. Coincident is the evolution of new marketing skills in data mining, web development, and "human-centric" applications awareness.

---

[2] radar.oreilly.com/2013/09/demographics-are-dead-the-new-technical-face-of-marketing.html.

Pricing

In its fundamental terms, pricing is the process of determining what consideration a company will receive in exchange for its products or services. It is one of the four Ps of marketing. The others include Product, Promotion, and Place. The key influences on pricing include a bottom-up one driven by manufacturing and distribution costs, brand, product quality, and competition. The second consideration is one of "what the market will bear" and rewards the most efficient producer and distribution channels. The influence of off-shore manufacturing has influenced this equation significantly. Added to this is the presence of Walmart-type "big box" distribution and Internet-based Amazon-like services that allow products to be available to consumers in almost real time. The customer can certainly "vote with his or her feet" and influence the demographic profile in a ways that were simply not possible earlier.

Warren Buffett's take on pricing is this, "Price is what you pay and value is what you get."[3] He further comments that "if you have the power to raise prices without losing business to a competitor, you've a very good business."

To determine proper pricing requires a sound strategy that embraces the market and its competitive forces, as well as the corporate requirements for profits and returns on their investments. There are three dominant forces that help determine a strategic direction for pricing, discussed next.

Cost-Based Pricing

This method is based on accounting data and focuses on a stated ROI corporate goal. It also embraces the concepts of breakeven (BE) and experience curve (EC) impact. Let's look at both.

Break*even* accumulates the fixed and variable costs in manufacturing a product or service and projects future trends of each based on the units produced. It then overlays the revenue generated by sales and further defines the quantity (units) required to have the revenue line exceed the combined costs. Interestingly, the slope of the revenue line is based on the average selling price of the product. From that point, the process generates profits (and cash). Before that point, it absorbs loss. The relationship is expressed as follows:

$$Breakeven = Fixed\ Costs\ /\ Unit\ Price - VariableUnit\ Cost$$

[3]Cited in *Technology Ventures* by Byer and Dorf, McGraw Hill, Third Edition, 2011.

Figure 7-4 shows this graphically.

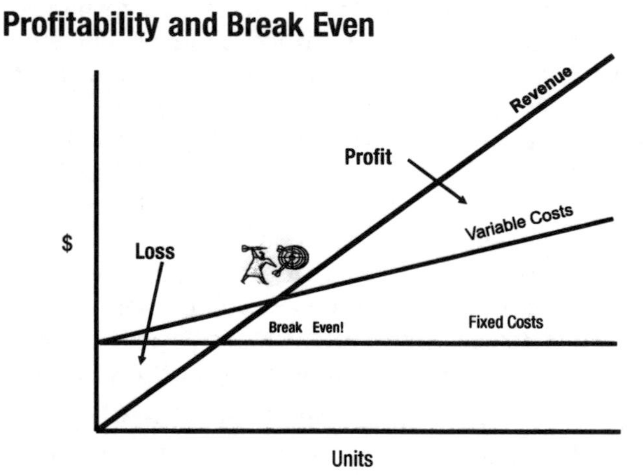

Figure 7-4. Profitability and breakeven analysis

At a certain point, the fixed and variable costs per unit are matched by the revenue generated from the sale of the product (or services). From that point onward, the production yields a profit to be enjoyed by the organization.

The ***experience curve*** is based on an idea developed in the mid-1960s by the Boston Consulting Group (BCG). It developed the idea that the higher the volume of experience a firm has in producing a particular product (or service), the lower its costs should be. Bruce Henderson, the founder of BCG and the author of the text entitled "Management Ideas and Gurus" (Economist/ Profile Publishing, September 14, 2009), claims that with each cumulative doubling of experience, costs decline by 23–30%.[4] In theory, this would imply a significant competitive advantage over time, as the company can control the pricing.

Time has altered the value of this model, as it does not incorporate the impact of innovation and change or off-shore competition. But as unit volume increases, automation and constant improvement can certainly sustain the value of experience.

Competition-Based Pricing

In certain product areas, price becomes the dominant criteria for procurement decisions. Retail segments such as automobiles and airlines are examples. Significant attention is focused on sales (for example, President Day sales

[4]*Management Ideas and Gurus* (Economist/Profile Publishing, 2008).

for automobiles), discounts, and bundled offerings (for example, "buy one get one free"). Although necessary, such marketing efforts tend to ignore demand cost functions and put corporate goals of profitably at risk. It also invites the creation of discount channels such as T.J.Maxx and Home Goods in the retail products market.

Competitive pricing is certainly one strategy to be considered. It has many alternatives, such as offering a family of products. General Motors manufactures (and sells) Chevrolet and Cadillac model cars. From a distance they are similar but can be sold at different price points to satisfy multiple markets. In reality, they are common in their ability to offer automotive transportation. Sometimes competitive pricing is used to move mature products in a way that extracts maximum value from declining profit margins.

Value–Based Pricing

Value-based pricing uses data based on the customer perception of value. A simple metric of cost-benefit analysis is used by many consumers.

A company located in the Boston suburbs called Phoenix Controls designed, sold, and manufactured a line of airflow controls used in harsh chemical environments. It worked on the principle of controlling regulated air flows in laboratories accurately and more linearly to demand. The basis for purchasing the equipment was manifold. It saved energy costs by precisely controlling the amount of flow needed in the workplace. It made the environment more energy efficient, quiet, and more temperature controlled. Each of these elements was factored into the proposal on a perceived cost- benefit basis as the buyer might perceive them. There were variations based on local conditions such as the local price of energy or labor. This method yielded a higher selling price than conventional cost-based models.

This value-based pricing has the highest impact on new or complex products where market or consumer procurement trends have not been identified. It also yields higher prices and gross margins. Those results help defer initial launch and development costs that are necessarily associated with new products. Because of this, it also invites competition earlier than a normal product cycle would.

One aspect of pricing that has changed remarkably is the influence of the Internet. An example is airline ticket pricing. Entrants like Travelocity and Expedia have entered the game and can produce a list of base flight costs between given cities for a large of array of airlines with the simple click of a button. The comparison is bit skewed by additional charges that can easily be added to the stated base number. Consumers have proven quite nimble in working around the limitations. This type of comparison can be done to almost any commodity that we all use. Pricing differences become invaluable and the strategic implications are enormous. Some of the tangible aspects include:

- Help the parent organization achieve its financial goals.
- Help position products (or service offerings) to attract customers.
- Help create a family of products (mix) to serve different customer requirements.
- Affects distribution channels that the products utilize to get to the customers.
- Can be used to infer product quality or level of utility.
- Can be used to support advertising and product discount strategies.

Pricing is possibly the most fluid parameter of the commercial cycle. It allows entry into a crowded market dominated by established players. It allows product (or service) offerings to be positioned in perceived value spaces in the market. It can significantly affect corporate profits. Given its importance, it is strange to see that pricing strategy is often far from the analytical or scientific disciplines. For anecdotal evidence of this, look at the Sunday flyer section of the newspapers. You'll see seemingly endless sales and discounts designed to entice customers to the retail shops.

Pricing strategy can be categorized in the following ways. This not meant to be an inclusive list but rather a sample of the many directions available:

- Absorption pricing. A method of recovering fixed and variable costs.
- Contribution margin-based pricing. Where the factors driving gross margin contribution are used to determine the price. Responds best to high volume but is somewhat insensitive to market conditions.
- Skimming. With new products market share is sometimes compromised to yield early recovery of costs. This is risky in terms of branding and inviting competitors.
- Decoy pricing. Multiple offerings of similar products are offered at different prices.
- Freemium. Free or introductory offerings are made followed by a subscription. Used many times in the software market to entice new customers.

- Loss leader. A product (or service) is introduced at a lower price with the resulting sacrifice of profits. Its goal is to attract new customers early.
- Market-oriented pricing. This competitive strategy reaches for market share and is dependent on what others offer their products for. Although chasing the market might be positive, the risk to profitability is large and relies on good manufacturing and distribution controls.
- Value-based pricing. Used with some markets where actual material cost is a small part of the overall cost structure. A software CD is an example. Its value is in the content and requires constant understanding of the marketplace to implement.

Competition

The arenas of commercialization, pricing, and competition are intertwined. In the literature surrounding technology commercialization, the role of competition in the evolution of a commercial product is complete with supporters and detractors. Even the early economic commentators had their say. Karl Marx, in his classic tome entitled *Capital*,[5] wrote that competition "impeded the free movement of capital" and allowed the "disruption of capital that had already been invested." Considering his criticism of capitalism, this might have been a valid perspective.

Adam Smith, the economist and a social commentator, in his much-touted *Wealth of Nations*,[6] spoke of the competing forces of self-interest and competition, while not acting with the intent of serving others, acted as an "invisible hand" that eventually served the benefit of mankind.

Joseph Schumpeter—a 20th Century economic and political thinker—takes a more positive view and argues that competition is not an efficient economic driver but does acknowledge that entrepreneurial innovation serves as an engine for growth. He points out that investment in R&D correlates to productive growth. Innovation, he argues, leads to "creative destruction" of old inventories, ideas, technologies, and skills. He outlines these theories in the classic book entitled *Capitalism, Socialism and Democracy*.[7]

Whatever import you place on competitive forces, it is clear that going forward, you need an applied strategy to learn from them and to deal with them.

---

[5]University of Bergano Press, third edition, 1999.
[6]Oxford University Press, 1776.
[7]Harvard University Press, 1934.

Competition and Commercialization

A modern context for competition allows for disruptive forces to be acknowledged and mitigated. The forces become a relevant fact of life and must be accommodated in the business or project plan. From a societal point of view there are three vital functions that competition serves. They include:

- Discovery. The discovery of new knowledge is an inherent attribute of human nature. Instilling this into the commercialization process yields a distinct advantage. It is well within the realm of self-interest. Sometimes it can be motivated by societal initiatives, such as the technological improvement that resulted from our quest for a clean environment. An example is the use of catalytic mufflers in automotive applications. The catalysis of exhaust gas had been known in the laboratory but it wasn't until the government standards for emissions were implemented that automotive muffler applications became standard.
- Selection and coordination. The actual procurement decision that transfers financial consideration for goods or services is a powerful form of "voting" for orderly innovative commercialization. Innovation ranges from ideas to breakthrough changes. It is the decision to buy that innovation that sets up an orderly progression of ideas.
- Control of power. The power to create wealth from innovation is controlled by how well the market presents and delivers the ideas to those that will eventually use them. It acts a force to distribute wealth. External control such as government regulations and societal rules also prevail. Antitrust is an example of such an artificial check and balance. Although those laws restrict the movement of ideas, they also even the potential abusive use of those forces toward commercial realty.

Competitors

Quantifying a project's competitive forces comprises several steps. They include:

- Identifying who the competitors are. They can be a totally new business, a new product, or even a new technology. They can be identified well before business opportunities are lost. Simple observations like their advertising, presence at trade shows, Internet sightings, patent searches, and simple dialogue with existing customers.

- Identifying them is not enough. Knowing what products they offer, how they distribute them, their annual reports (if public) or their SEC filings, their pricing schedules, who they identify as customers, and possible their financial resources. Checking their literature is a great starting point.
- Dealing with them. Competitors offer an opportunity to learn and innovate. Observing how they go to market is a chance to improve that process. Sometimes the best strategy is to build on your own strength. Improving your customer service response function is an example. Continuous internal improvement is solid way to improve customer effectiveness and competiveness.

The Impact of Competition on Marketing

As economists debate the intersection of competition and innovation, it becomes important to focus on the impact of this debate on the marketing of new products and services in the commercialization cycle. Sometimes it is positively disruptive. Examples such as 3M's Post-it Notes and Toyota's Prius created new market segments. Other times, this power of innovation is disruptive in a negative sense.

In Clayton Christensen's text titled *The Innovator's Dilemma,*[8] he explains how innovative changes in the size, form factor, and performance of disk drives occurred so quickly that the current offerings were obsolete before they had a chance to exploit their full product lifecycles. Subsequently, there was a fallout of disk drive producers from 35 to 14 companies in the industry. Innovative changes also tend to provide downward pressure on pricing.

Competitive forces exist in any viable commercial opportunity. Something will be challenged or face change when an innovative product or service is introduced. Whether perceived as a positive or negative force, they are an inevitable fact of life. It seems analogous to strategic selling (cited earlier in the book; see Chapter 5), where many influences surround the actual purchasing decision process. The influences could be financial (and cost), technology, market-driven items such as branding, quality, or exposition of features and benefits. Like the strategic selling analogy, any one (or combination of) force can compromise a project's effectiveness, and have the power to negate a sale.

---

[8]New York: Harper Collins, 2003.

It is the CEO who must correctly identify the competitive risks, then act to offer mitigation alternatives to overcome them. Once identified, there are usually multiple options to neutralize the threats. In product design, for example, there are families of product design that can be populated to allow multiple technical personalities from a common platform. Anticipating them and providing options is where the successful projects thrive. In the marketing area, there are multiple branding, distribution, and pricing strategies for any one product or service offering. Within the marketing space there are still fundamental tools such as SWOT analyses, five-point strategy tools, and many more that help outline available alternatives.

Going to Market

One more time in our search for improved paths to commercialization the need for succinct planning arises. This time the issues are quite tactical. It starts with a situational analysis that describes the customer behavior and the market environment as it currently exists. It outlines the outside forces that can impact the success of the project. These include anything from competitive elements to regulatory ones. Clearly, they define a target customer audience and describe their purchasing patterns (and quirks). Most important they offer a strategy or approach as to how to overcome them.

New products and services have the potential to disrupt and alter the brand and image of the parent company. Most times this direction is positive but its impact needs to be studied. The nature of the marketing disciplines certainly differ with a company's existence and the lifecycle of its operations. The planning disciplines are certainly more prevalent and formal as the organization matures. Finally, the budgets that control cost and the resulting impact on upside revenues need to be delineated.

Why bother to embrace the effort to create this plan? The answer lies in the fact that most new projects seem to stumble in market domain. The risk is that the company expends resources but does not benefit from the revenues and resulting profits it hopes to achieve. This can be avoided.

Why Marketing: A Summary

Of all functional disciplines, marketing probably has the most significant impact on the success of the outcome. The results of a successful marketing and sales campaign translate directly into financial performance and overall project satisfaction. Technological components, human resources, operational issues, and even financial support can usually be corrected. Marketing issues and their resolution are the litmus tests for the customer and their quirks. These issues can't be solved in real time and thus must be thought through and planned.

All issues surrounding new products and services destined to commercialization must face the scrutiny of being quantified. It is the only hope of rationalizing those elements that they can be reduced to analysis. It turns out that there are many parts that can be quantified. Let's look at some of the most important elements in the next chapter.

CHAPTER

8

The Numbers

In early stage companies (or projects), there is a need for close control of capital resources. Yet even with staff and control systems in place, money seeps through the cracks, never to be seen again. Clearly, better tools are required to navigate through this quandary. I present a financial model that not only helps control the variability of cash flows, but also allows you to change the assumptions behind the numbers to give it a predictive capability. The change in the accounting and finance functions at each stage of growth will also be presented. They are dramatic and affect each function in the company.

The world of numbers completely embraces the world of commercialization decisions. The value of numbers is that it allows us to measure and compare our performance against larger goals such as corporate return investments and periodic (monthly) goals. It certainly flags issues and allows us the calculus of repairing them. Numbers also allow us to choose between alternatives in such a way that compares various attributes (ratios) in multiple projects. Finally, numbers allow the outside world to observe the performance of projects and activities. Let's look at each category in more detail.

The Internal Space

In the path of commercialization decisions, there is one particular numerical gate for decisions. It is the Return on Investment (ROI). In the simplest terms, it is the calculation of the amount of capital invested in a project and its anticipated returns. If some minimum value cannot be obtained, that becomes a basis for rejecting a project. You read about ROI as a feasibility analysis decision point in Chapter 5. The calculation is as follows:

$$ROI = (Gains - Investment\ Costs) / Investment\ Costs$$

There may be mitigating circumstances where an idea is brought forward with a lower than anticipated return, perhaps because it provides the parent organization with a strategic entry to a new market segment. This type of project is sometimes called a "loss leader." The decision is strategic and can override any financial considerations.

ROI alone does not allow an absolute financial decision about a project to be made. There may be multiple projects that promise attractive returns, but in total would put a significant drain on available capital from the parent organization. A means of calculating the rates of return among projects is needed. It is called the Internal Rate of Return (IRR) and quantifies the rate that capital expenditures will be consumed. It also allows you to compare the parent organization's ability to raise and provide adequate project funding. There are variations available such as adding investment capital, loans and selling equity (stock) in public markets that can add to the financial options available. These options are commonly used but not without consideration of overall equity dilutions and default penalties in debt financing.

Another financial consideration flows from the perceived strength of the parent organization as measured by its balance sheet (B/S) and its embedded ratios. This document is usually compiled on a monthly basis and represents a presentation of the assets and liabilities of the parent corporation. Within its structure, the cash and tangible assets are shown. In addition, it presents the value of obligation against assets such as loans and accounts payable. It is one of three controlling financial documents. The Income Statement and Statement of Cash Flow are the other two and will be covered later in this chapter. A sample balance sheet (sans numbers) is shown in Figure 8-1.

Sample Balance Sheet

XYZ Company

Month, Year

Assets	
Current Assets	
Cash and Securities Accounts Receivables Inventory Prepaid Expenses Other	
Total Current Assets	
Other Assets Goodwill Intellectual Property	
Total Other Assets	
Total Assets	

Liabilities	
Current Liabilities	
Accounts Payable Interest Due Short Term Notes Accrued Expenses Total Current Liabilities	
Long Term Liabilities	
Long Term Debts Mortgages Total Long Term Liabilities	
Owner's Equity	

Figure 8-1. Sample balance sheet

The balance sheet presents a listing of all assets and liabilities of the company. The format for this is the same whether it is General Motors or a startup company. They are generally delineated in two sections of current (less than 30 days) and long-term (beyond 30 days) entries. The first assets look at cash, inventories, and accounts receivable (money owed to the company). These are the first items that suggest how much capacity the company has to invest in new commercialization projects. Outside sources such as banks consider assets like inventories and receivables (on a discounted basis) as viable collateral to loans, but that is only one part of the story.

The assets are offset by a listing of the organization's liabilities. Top ones include payables (such as money owed to vendors) and others like outstanding loans and mortgages on properties. If there are many debt obligations presented, the financial judgment is that the organization is highly "leveraged" and thus vulnerable to risks. Included in this is the judgment of whether the company is ready to absorb the risk of new commercial ventures.

There is an important ratio of current assets to current liabilities called the "current ratio." A positive number suggests that the organization is capable of absorbing financial risks and responsibilities. It also deals with the question raised during the feasibility analysis as to whether the company is ready for supporting new commercialization activity.

Another consideration is the fact that all commercialization projects are not the same with respect to their timing of both investments and anticipated returns. This is particularly of interest when decisions between project alternatives need to be considered. There is an analytical financial model called the Net Present Value (NPV) that rationalizes time-based components to a present number. NPV allows investments made over periods of time to be mathematically brought to a current value so that it may be compared with other projects being considered. It is particularly effective in simple, linear projects. Projects with large variations in capital investment require more complex analytical tools.

We soon see that there are multiple measurements to characterize and anticipate the capital requirements of a given project. Within the enthusiasm of quantification comes a quick caution about "analysis paralysis." There may be significant numerical arguments for or against a given project. Many profound decisions have been made on a much more balanced and intuitive "gut" feel. Examples are many and quickly allow us to dwell on the Jobs transformation of Apple. Many good decisions are made on an irrational basis. It's a bit like a pendulum that swings from both sides and eventually finds an equilibrium that allows both sides to prosper. Before too much effort is focused on finding a balance, we should look at the project's internal elements.

The Internal Balance

In an earlier chapter we looked at planning and its multiple benefits to the commercialization process. It became obvious that one outcome of the process would be an indication of the capital requirements to fund it. Whether a formal PERT or Gantt methodology is employed, it yields expenditures and the timing of them. Many times the numerical output is also a basis of a budget process. The detail and sophistication of the budget varies with the complexity of the project and the sophistication of the parent company's internal processes.

It is worthwhile to review the penalty of improperly scoping the numbers. To list a few:

- Missed market window opportunity. Insufficient budgets means improper product development funding or market launch efforts.
- Shoddy technology (not proved or finished). The time and human effort to properly develop technology and test it adequately is somewhat tragic if not realized. It may indeed be a novel or innovative idea, but its improper gestation time for development can compromise its effective space in commercialization. Not only are opportunities missed, but recovery is expensive in terms of overtime and premium delivery rates.
- Staff burnout due to underfunding. Early stage teams tend to be understaffed and the demands on their precious resource of time are overburdened. Driven by inadequate budgets and timelines, it's fully understood. But the impact on the effectiveness of the delivery is negative. Shortcuts are taken and people simply leave due to the stress. If unrealistic planning and investments are the cause, it becomes one more reason to ensure proper planning and budgets.
- Improper sales/marketing support in place. With the overall scope and speed of the Internet and its phenomenal ability to transmit ideas and products around the world, the commitment to finding new and appropriate channels requires new skills and perspectives. If that requirement alone weren't important, the ability to recast (change) and adapt ideas moves at the same lightning speed. I recently went to a dealership to shop for a new car. When I asked the salesman for a written brochure, he said they don't print those anymore and that I should look at the company web page. Old marketing budgets will no longer serve as a benchmark for going forward.

- Simple overall project failure. Even with the most innovative and adaptive thinking about new projects, there is still a potential element of failure. Many times there are external events such as governmental regulation changes and new technologies of products that present unforeseen obstacles. New descriptors of successful projects include words such as "nimble," "adaptive," and even "learning" organizations.

In common, the budget formats allow an expression of timing. This is particularly important when funding long lead items such as capital equipment or new physical plant expenditures. Breaking out and committing to long lead items prevents the havoc of scrambling for resources late in the project. Certain project planning tools have the capability to sort out those times automatically. Compounding this is the inevitable overtime labor charges, FedEx premium delivery fees, and vendor penalties. Not to mention the ill will and missed market opportunity windows that come with all this.

Budgeting

In a recent Google search of the word "budget," I received over 11 million hits. There are many uses of budgets and formats, as well as numbers of suppliers and vendors supporting them. I have been a fan of budget formats that are a) created for a particular use such as commercialization projects and b) that are easily transportable to the overall financial instruments utilized by the parent organization. This is accomplished by adapting common categories (buckets) and adaptation rules. In the end the use of numerical tools must integrate into the overall financial structure. To see this, let's first look at the income statement shown in Figure 8-2.

Sample Income Statement
Sheet
XYZ Company
Month, Year

Income		%
Revenue		
Less: Cost of Goods Sold (COGS) Materials Labor Mfg. Overhead		
Total COGS		
Gross Margin		
Sales, Generals and Admin. (SGA) Sales and Mkt. R & D Admin		
Total S, G and A		
EBITA		

Figure 8-2. Sample income statement

The income statement (I/S) is a compilation of the revenue that is generated and can be generated by a proposed project. It then applies the expenses to be charged against it. Those expenses are delineated in two categories of fixed and variable elements. The statement is a quantification of the breakeven graphic presented earlier in the text. The graphic is a quantified version that shows us where the profitability of the project lies. It is sometimes called the profit and loss statement. If it's done in the context of the rules and format of the overall cooperate entity, it allows easy transport to the overall financial documents later.

It can also be projected into future time periods. The accountants refer to these forward-looking projections as Pro Forma documents. The term is derived from the Latin "as a matter of form" or "for the sake of form." As a projection it is not held to the accounting standards (Generally Accepted Accounting Practices—GAAP) and thus may not include certain recurring and exceptional (restructuring) charges. It does, however, provide us with insight as to how the project will fare in the future. It is also a useful tool to compare this project to others through certain ratio analyses. We'll look at those later. Let's first drill down into the document itself.

Revenue

Revenue is the financial capture of events that are sold to others. In the transaction of selling, there is a simple exchange of products and services for financial "consideration." Consideration is a collective term that refers to cash, cash equivalents, or promise such as credit allowances. The process of selling is controlled by the Uniform Commercial Code (UCC) and presents the opportunity for revenue "recognition" at the time of the sale. That last part of revenue recognition is a bit tricky and focuses on long-term leases or long-term payment plans. The challenging part is usually distorted by other considerations like tax liabilities or offsetting payments obligations.

In the category of revenue there is another source of capture called "passive" revenue. It is the recognition of sources that are not predicated on a selling transaction. Examples include rental income and patent royalties. In common, all sources combined (called the total revenue) offer a basis for a powerful comparison to other companies and projects. In ratio analysis, the revenue becomes the denominator at 100% so that R&D, for example, can be presented as a percentage of sales. This allows one to compare different accounting periods and organizations. It becomes the basis for monitoring trends over time. This is particularly important in tracking a project's progress toward commercialization. In many comparisons, it is as simple as adding a column of ratios as the percentage of sales on the income statement.

Those Ratios

A traditional income statement starts with a presentation of the variable costs of the operation. They are referred to as the cost of goods sold (COGS). Within this category are three conventional buckets. They are Materials, Labor, and Manufacturing Overhead.

If you drill down into the first of these, you see the challenges to management appear. In small volumes, unit material costs are generally higher. Purchasing a part at quantities of 10 to 100 is less efficient than purchasing it in thousands. Labor offers analogous savings as automation and robotics enter the picture.

As the total cost of goods is subtracted from the total revenue, it defines a term called Gross Margin (GM). It is a first measure of the manufacturing efficiency of the company or project under consideration. It allows one to compare competitive projects and see any relative measures of improvement in existing ones.

It also opens up an option of being increasingly utilized. It is the possibility of outsourcing production to others and offshore. We all know of the trend to make goods in the Far East. What triggers this is the inability to compete in gross margins or manufacturing efficiency. This concept reaches into service industries. There we see increasing trends in accounting, software development, and medical record processing being outsourced to India and other places in the world where the labor rate is lower.

Some Are "Fixed"

As we saw in the breakeven analysis, the next category of costs delineated in the Income Statement are fixed costs. They are captured in the accountants called the Sales, General, and Administrative (S, G, and A). These arcane accounting terms embrace the areas of Sales, Marketing, Research and Development (Engineering), and Administrative areas such as Human Resources, Accounting, and Legal. A robust gross margin allows for increased amounts of SGA expenditures. Two major areas are in focus as they benefit future activities.

First are the sales and marketing areas. The value of a significant marketing expenditure allows a project to benefit from advertising, trade show exposure, and adequate support literature. In 1971, Gillette, the personal goods manufacturer located near Boston, announced a relativity new shaving concept and product in its Trac II razor. In design it appeared to be backward from the traditional double-edged technology. It suffered low acceptance. Gillette ran a two-page advertisement in *Life* magazine in December of 1971 with only incremental results. In a somewhat bold marketing move, they decided to do a mail campaign giving potential consumers a free razor handle and two blades. The results were immediate; today Gillette's lines of personal products are considered the standard. Adequate marketing funding reserves were sufficient to turn a somewhat soft market response into a positive outcome.

Having adequate sales effort certainly follows. In a telling article entitled "The New Science of Sales Force Productivity," published in the *Harvard Business Review*,[1] Dianne Ledingham, Mark Kovac, and Heidi Locke Simon of Bain and Co. dispelled certain myths about sales force productivity. The first was that simply adding more sales representatives was not effective. Also, they found that relying on superstars (referred to as "rainmakers") was not as effective as methodically utilizing a broad range of skills. They demonstrated in five-year sale productivity model that consistently improving the sales force productivity yielded a measure called Sales per Rep productivity of $4.1 million sales per rep, versus a $3.2. That increase was 17% better than a peer group competitor. Measuring ratios against competitors is common practice. It is called "comp" (short for comparable) analysis and considered effective. This is certainly true of more common ratios like sales per revenue, which can also be compared to public companies where the data is easily accessible.

Analogous arguments can be made for R&D expenditures. In earlier times, corporate R&D was characterized by large basic research facilities such as RCA's Sarnoff Labs, Bell Labs, Homer labs (Bethlehem Steel), and Ford-Philco. R&D had a very tangible presence. I suspect that the commercial advantage of basic technology was less fruitful than anticipated. In addition, government expenditures in university-based basic research continued.

---

[1] `www.HBR.org`, September, 2006.

Development continued. Certain industries such as biotech, materials, and semiconductors, for example still pursue a mixture of research and development because of the nature of their business. All of this combines to reward internal commercialization activity for future financial opportunities. One aspect of these trends is that commercial opportunities will become shorter in vision and probably result in more collaborative opportunities and more applied technology. Commercial opportunities will change in that they must prove themselves positive cash generators sooner. The ability to sustain long-term risk has diminished.

There is a broad category, called "other" fixed costs, that indirectly affects this discussion. It is the other administrative costs. They include accounting, legal, utilities, insurance, and so on. The indirect aspect is that they are always under scrutiny to be kept lower. With the breakeven analysis in mind, it becomes apparent that, if fixed costs are lowered, the potential for earlier profits encourages more commercial activity. With this in mind, we now look at results in financial terms.

The Results

When all the fixed and variable costs/expenses are charged to the revenue of the income statement, the remainder is referred to as the "operating income."

Expenses that are variable change with respect to the manufacturing of a product or services. With each unit made, there is a charge for material and labor utilized in manufacturing. The same holds true for services where professional service hours might be substituted for materials and labor. Finally there is overhead charge for the sources and utilities utilized.

The more formal descriptor utilized by accountants for operating income is Earnings Before Interest, Taxes, and Amortization (EBITA). The term recognizes all the elements that come from producing and selling products and services. It allows one to compare projects and companies. The passive components of interest, amortization, and tax obligations are then subtracted from the EBITA to yield the net income for the organization. Included are the tax obligations, net interest, and depreciation.

These passive (non-operating) components of expense can be significant and might be as much as 10% of the stated revenue. Some elements of EBITA are quite strategic. One example is tax obligations. Certain expenses and investments can be sheltered by offsetting tax incentives. R&D tax credits can be applied against R&D expenditures and even be carried forward to future accounting periods. This incentive may sway a parent organization to make that investment. Similar arguments might be made for job creation and even location incentives. Loss carry-forward formulas may even allow tolerance for slower profit growth considerations. The influence and impact of these tax incentives can be significant and this can be factored into the decision processes. These considerations rarely drive the commercialization processes with same intensity and priority as innovation, market share, and growth, which is the way it should be!

Another aspect of the EBITA to net income calculations are loans and the interest expenses they carry. It is seductive to consider the non-dilutive aspect of loans, but the obligations they carry are counter to the operational instincts of most business decisions. To some, like me, debt obligations are an anathema of sound financial decisions. They exist beyond the use of the money and the interest remains fixed on the books. Given that bias, loans can extend a given corporation's reach into new products, services, and markets.

As with so many business decisions, a balance is required. The implications of balance sheet ratios may finally dictate debt considerations. Many times the amount of debt an organization can incur is dictated by the lending sources such as banks. It takes the form of restrictive covenants that may allow the lender to call the loan if it's not met. Ratios such current ratios of assets to liabilities are used frequently. If a balance sheet does not indicate a strong enough equity position to absorb debt, these ratios can be adversely affected. It is referred to as the liquidation potential of the firm. Lenders are not in the position to absorb the risk of a project or a company's future. Risk investments are the role of the equity. Stock offerings are common alternative vehicles to raise the capital needed to fund new projects. That calculation is based on the premise that there will be sufficient future increases in earnings and resultant increases in perceived value proposition to warrant them.

Within this space, before net income, is a place for "other" passive considerations. An increasingly visible one is the impact of regulatory incentives and obligations. The long litany of government interventions such as EPA, SEC, OHSA, and IRS were once not considered material. Today, their magnitude alone dictates that they be given both visibility and strategic considerations. The sum of these items is subtracted from EBITA and results in negative differences in net income.

In a sense, net income is the final arbitrator of the effectiveness of the corporate operating entity. Successful new commercial activity contributes to this and to its growth. Outside evaluation of the corporations rely on this number when the critical ratio of its stock price is compared to its earning on price per share/earnings per share calculation. This is particularly true of companies whose stock shares (equity) are sold in public arenas such as the NY Stock Exchange (or equivalent).

Modeling This

So far the dialogue about the numbers focuses on the importance of each of the financial components and how they relate to the project. The number-link performance (both anticipated and realized) to the corporate environment and numerical attributes of the world of opportunity that the project addresses. Yet, it is known that most projects fail because of lack of cash. In the search for tools that allow us to better predict outcomes, we look at the financial models of the proposed projects.

In my early days of teaching entrepreneurship at WPI, I developed a multilevel model that has proven useful in the analysis of both funded projects and in the classroom. The model mimics the breakeven model cited previously and estimates the balance sheet capital to present first order estimates of the capital required to fund a project.

The model acknowledges the ambiguity of underlying assumptions and offers a process to avoid their implied variability through a series of controlled steps. Delineating the controlling assumptions serves to isolate them for analysis and allows the opportunity to dynamically update them as information quality increases. It also offers a dialogue tool to assimilate multiple inputs of expensive items such as labor manloading and analysis. The model is shown as Figure 8-3.

Template Income Statement (Single Product)

	Month 1	Month 2	Month 3	Month 4	...	Month 12	Total	%
Number of Units	1000	1	1	1	...	1	1011	
Average Selling Price per Unit	$35	$35	$35	$35	...	$35		
Revenue	$35,000	$35	$35	$35	...	$35	$35,385	0%
Less: discounts	$3,500	$4	$4	$4	...	$4	$3,539	
Total Revenue	$31,500	$32	$32	$32	...	$32	$31,847	
Variable Costs (COGS)								
Panel								
Materials	$11,000	$11	$11	$11	...	$11	$11,121	
Labor	$13,000	$13,000	$13,000	$13,000	...	$13,000	$156,000	
Overhead	$48	$48	$48	$48	...	$48	$573	
Total COGS	$24,048	$13,059	$13,059	$13,059	...	$13,059	$167,694	
Gross Margin Contribution	**$7,452**	**-$13,027**	**-$13,027**	**-$13,027**	...	**-$13,027**	**-$132,309**	-415%
GM (%)	21%	-37221%	-37221%	-37221%	...	-37221%	-374%	
Fixed Costs (SGA)								
Sales/ Marketing								
Sales Salaries	$0	$0	$0	$0	...	$29,792	$126,208	
Sales Expenses	$1	$1	$1	$1	...	$543	$3,804	
Marketing Salaries	$0	$0	$0	$8,125	...	$8,125	$73,125	
Mkt Expenses	$2	$2	$2	$2	...	$501	$2,020	
Total Sls/ Mkt	$3	$3	$3	$8,128	...	$38,960	$205,157	644%
R& D								
R & D Salaries	$13,867	$13,867	$13,867	$19,283	...	$24,700	$247,650	
R & D Expenses	$501	$501	$501	$501	...	$501	$6,012	
R & D Other	$1	$1	$1	$1	...	$1	$12	
Total R & D	$14,369	$14,369	$14,369	$19,785	...	$25,202	$253,674	797%
Admin								
Admin Salaries	$813	$813	$813	$813	...	$9,479	$43,875	
Admin Expenses	$500	$500	$500	$500	...	$500	$6,000	
Consultants	$1	$1	$1	$1	...	$1		
Rent (Office)	$380	$380	$380	$380	...	$380	$4,560	
Utilities (Office)	$150	$150	$150	$150	...	$150	$1,800	
Insurance	$200	$200	$200	$200	...	$200	$2,400	
Other (Legal, Acct)	$2,000	$2,000	$2,000	$2,000	...	$2,000	$24,000	
Warranty	$1	$1	$1	$1	...	$1	$12	
total Admin	$4,045	$4,045	$4,045	$4,045	...	$12,711	$82,659	260%
Total SGA & A	$18,416	$18,416	$18,416	$31,958	...	$76,874	$541,490	
EBITA (Operating Income)	-$10,964	-$31,443	-$31,443	-$44,985	...	-$89,901	-$677,337	-2127%
Provision for Taxes, Interest	$0	$0	$0	$0	...	$0		
Net Income (Loss)	-$10,964	-$31,443	-$31,443	-$44,985	...	-$89,901		
Cumulative Income (Loss)	-$10,964	-$42,407	-$73,851	-$118,836	...	-$677,337	-$677,337	-2127%
Capital Investment								
Equipment	$15,000	$15,000	$15,000	$15,000	...	$15,000		
Inventory	$10	$10	$10	$10	...	$10		
Working Capital	$10	$10	$10	$10	...	$10		
Total CapEx Investment	$15,020	$15,020	$15,020	$15,020	...	$15,020		
Capital required	$25,984	$57,427	$88,871	$133,856	...	$692,357		

Figure 8-3. Income statement template

Note To download a copy of the income statement template, go to www.apress.com/9781430263524 and click on the source code tab.

In the planning stage of a project, there is significant uncertainty and variability in the supporting numbers. Professional salaries jump out quickly because of their magnitude.

An approach used in this model, for example, would be to look at a given functional salary line in a supporting manloading per month. This would reveal the anticipated timing of hiring plans. A look-up table of salaries and benefits could then be applied as a multiplier to the results of the loading schedule (Figure 8-4). This enables a constructive conclusion about the magnitude of salary expense (an assumption), without the concern of how well the individual cells were loaded with the distribution of the numbers.

Man Loading

	Month 1	Month 2	Month 3	...	Month 12				
Labor							30%		
Technician 1 (Panel)	1	1	1	...	1	$50,000	$15,000	$65,000	$5,417
Technician 2 (Panel)	1	1	1	...	1	$50,000	$15,000	$65,000	$5,417
Technician 3 (single test)	0	0	0	...	0	$50,000	$15,000	$65,000	$5,417
Admin	1	1	1	...	1	$20,000	$6,000	$26,000	$2,167
Sales/ Marketing									
Sales Manager	0	0	0	...	1	$100,000	$30,000	$130,000	$10,833
Inside Sales Support	0	0	0	...	1	$75,000	$22,500	$97,500	$8,125
Salesperson	0	0	0	...	1	$80,000	$24,000	$104,000	$8,667
Sales Admin	0	0	0	...	1	$20,000	$6,000	$26,000	$2,167
Market Manager	0	0	0	...	1	$75,000	$22,500	$97,500	$8,125
Associate	0	0	0	...	0	$50,000	$15,000	$65,000	$5,417
Mkt Admin	0	0	0	...	0	$30,000	$9,000	$39,000	$3,250
R & D									
Ch Scien, Officer	0.8	0.8	0.8	...	0.8	$160,000	$48,000	$208,000	$17,333
Lab Tech 1	0	0	0	...	1	$50,000	$15,000	$65,000	$5,417
Lab Tech 2	0	0	0	...	1	$50,000	$15,000	$65,000	$5,417
Admin									
Office Manager	0	0	0	...	1	$25,000	$7,500	$32,500	$2,708
bookkeeper	0.5	0.5	0.5	...	0.5	$15,000	$4,500	$19,500	$1,625
QA Manager	0	0	0	...	1	$35,000	$10,500	$45,500	$3,792
Shipping Person	0	0	0	...	1	$20,000	$6,000	$26,000	$2,167

Expenses

	Month 1	Month 2	Month 3	...	Month 12
Labor					
Technician 1	$5,417	$5,417	$5,417	...	$5,417
Technician 2	$5,417	$5,417	$5,417	...	$5,417
Technician 3	$0	$0	$0	...	$0
Admin	$2,167	$2,167	$2,167	...	$2,167
Total Labor	$13,000	$13,000	$13,000	...	$13,000
Sales/ Marketing					
Sales Manager	$0	$0	$0	...	$10,833
Inside Sales Support	$0	$0	$0	...	$8,125
Salesperson	$0	$0	$0	...	$8,667
Sales Admin	$0	$0	$0	...	$2,167
Sales Salaries	$0	$0	$0	...	$29,792
Travel	$0	$0	$0	...	$542
Commissions	$1	$1	$1	...	$1
Sales Expenses	$1	$1	$1	...	$543
Market Manager	$0	$0	$0	...	$8,125
Associate	$0	$0	$0	...	$0
Mkt Admin	$0	$0	$0	...	$0
Marketing Salaries	$0	$0	$0	...	$8,125
Trade Shows	$1	$1	$1	...	$1
Literature/ Mailings	$1	$1	$1	...	$500
Mkt Expenses	$2	$2	$2	...	$501
R & D					
Ch Scien, Officer	$13,867	$13,867	$13,867	...	$13,867
Lab Tech 1	$0	$0	$0	...	$5,417
Lab Tech 2	$0	$0	$0	...	$5,417
R & D Salaries	$13,867	$13,867	$13,867	...	$24,700
Supplies	$500	$500	$500	...	$500
Misc	$1	$1	$1	...	$1
R & D Expenses	$501	$501	$501	...	$501
R & D Other (equip)	$1	$1	$1	...	$1
Admin					
Office Manager	$0	$0	$0	...	$2,708
Bookkeeper	$813	$813	$813	...	$813
QA Manager	$0	$0	$0	...	$3,792
Shipping Person	$0	$0	$0	...	$2,167
Admin Salaries	$813	$813	$813	...	$9,479

Figure 8-4. Manloading and salary look-up table

It shows the bottom-line implications immediately. Utilizing formula relationships does this. The clarity of a salary look-up table allows dialogue about the overall salary more effectively than the mechanics of the model's distribution. As the model and its projections mature, the accuracy of the data increases and the overall predictive effectiveness of the model follows.

Note To download a copy of the manloading and salary look-up table, go to www.apress.com/9781430263524 and click on the source code tab.

The Outside View

There are multiple views of the numbers from outside the organization worth noting. Beyond the operational governmental agencies like the FDA, FAA, and others that control elements of quality, safety, and the delivery of project activity, there are several that focus on the numbers and their consequences. The first and most invasive one is the Internal Revenue Service (IRS). This Federal agency reports to the Department of Treasury and thus has cabinet level visibility. The IRS has the responsibility to collect taxes and administer the Internal Revenue Code. The Code contains the operating definitions of tax collection that allows the tax process to work. The dimensions of the Code are not trivial and are estimated to be 70,000 pages in length. The agency collects multiple forms of tax revenue and controls corporate taxes, individual taxes, employment taxes gift and estate taxes, and so on. It also administers the tax collection process and the myriad forms utilized by the agency. Since the Code even goes down to the level of codifying what is the definition of an employee, for example, care must be taken in implementing the Code's provisions. Although it can be done by individuals, it is best left to the domain of specialists and accountants.

Another dominant force on the structure of numbers on a commercial project is the Securities and Exchange Commission (SEC). The focus of the SEC is defining the securities and ownership of large public companies. It also sets the stage for how these securities will be administered. The SEC has three broad mandates. They include the protection of investors, maintaining orderly equity markets, and facilitating capital formation. They accomplish these mandates by an elaborate system of public reporting characterized by corporate annual reports. These are a structured reporting system prepared on an annual (and in some cases quarterly) basis.

The SEC also administers the regulations of the Sarbanes-Oxley and Dodd-Frank bills, which help establish the boundaries of the governance function of the board of directors. Finally, the SEC helps define the structure of equity investment in companies. Here is the basis for a discussion about the equity structure early stage companies. Some issues deal with passive or pass-through provisions of taxes, depreciation, and other expenses that can reduce an investor's personal tax obligation.

Finally, there is a placeholder for local government regulations. These deal mostly with filings and the number of investors allowed for tax and structural reasons. These provisions are administered on a state-by-state basis.

Summing Up

This "journey" into numbers and accounting serves to show the important impact they have on commercial activities. They reveal an organization's capacity to embark on new projects as well as the operating income effects of the development and implementation of new products and services. Beyond the metrics of performance, they also can be used to compare organizations and industry standards.

Every element of the parent organization's numbers is affected by a project's financial dimensions. This is true whether it is a call on resources like cash or an influx of resources based on the project's success.

Another dimension of the numbers analysis comes from recognition of where the company is in its lifecycle. Most of analysis of a project's metrics and their impact on the parent organization in an early stage entity can be performed on computer spreadsheet such as Microsoft's Excel product. Basic reporting and its repetitive transactions might be well executed on a store-bought "shrink wrap" package such as Intuit's QuickBooks or Peachtree Accounting (now called Sage 50). They are inexpensive and do not necessarily require significant experience in their execution. As the organization moves along the growth curve, additional financial responsibilities occur in terms of reporting and analysis. Various Federal and external requirements, internal reports to the Board and budget committees, and considerations of multiple currencies enter the discussion.

Finally, the reporting requirements of public entities created by the IRS, SEC, and now Sarbanes-Oxley (the law that controls the information flow for boards) enters the arena. Complex accounting programs capable of consolidating that information, such as SAP and Oracle, now enter the dialogue. They require additional sophistication to the folks who operate them. Management must be capable and sophisticated enough to absorb the information.

How new ideas are implemented becomes as important as the value of the ideas. Beyond the numbers, this is one of the reasons investors look at the team so closely. Their experience and skills represent the ability of the project to deliver its promised results. In the next chapter, we will look at various organizational models and compare the options for acquiring the talent and motivating them to achieve the goals of the project and organization.

CHAPTER

9

Organizational Dynamics

When you start a project, you need the right people functioning in the right structure and in the right place. Early stage companies must compete with the larger resource base of established organizations, making attracting talent more difficult. Then, even when you think you have the right people in place, you need to bring the best out of them given the culture and overall organization of the company or project team. This chapter will help readers focus on the OD issues that will best lead to long-range success.

Getting Organized

There are many elements that contribute to the probability of the success of any commercial entity. Possibly the most significant element is the team of individuals called on to actually execute the idea's plan and realize its commercial potential. Within the set of decisions that control the organizational model is the choice of framework within which they will operate. Many potential investors/bankers look at the team as the most significant element of their assessment. They are the ones who can deliver the project's promise. Other investors focus on financials and markets. Although domain expertise is important, the role of the "serial entrepreneur" cannot be diminished. Allegedly they bring perspective, experience, and industry recognition. Some argue that these attributes are overstated and value the naiveté and enthusiasm of the new entrepreneur.

In an article published in the April 2011 *Harvard Business Review* entitled "What Entrepreneurs Don't Learn from Failure," Deniz Ucgasaran, Paul Westhead, and Mike Wright cite a study in the UK in which 576 serial entrepreneurs were interviewed. The data showed that the outcomes of their current projects were not significantly better than those started by first-time entrepreneurs. In a pithy reported comment one person was quoted as saying "Spending your time thinking about what happened is a ticket to the graveyard." Investors who value experience highly also carry the responsibilities of exploring the "lessons learned" during an entrepreneur's previous exercises.

Rather than join in this discussion about the value of serial experience, I offer that the role of the team is not static and changes with each shift realized during a company's growth. Thus the skillsets presented by a close-knit startup team may not be sufficient to grow the company or its projects. The good news is that these changes can be anticipated and resources can be brought to bear to change that limitation. Examples include training and consulting. The goal is to provide resources that enable the team to make these transitions. This idea is suggested presented in Figure 9-1.

Stages of Growth – Functional Implications

	Pre-Investment	Early Stage	Series A	Subsequent
Management	Informal	Teams	Management Structure	Committees, Metrics
Accounting System	Excel	Quick books	ERM, CRM, MIS	Consolidated
Accounting People	You and me	P/T CFO	CFO (HBS?) (teams)	Treasurer
Sales Org	none	Focused sales by all	Teams, reps	Multinational
Sales People	See above	Senior mgt.	Sales manager	Professionals
Operations	none	All hands	Teams	Structure, procedures
Ops people	See above	First COO	Integrated mfg.	Outsourcing
Marketing	Primitive, assumptions	Verification	Formal, trade shows	Consultants, global
Marketing People	Advisors	Part time CMO	Trade shows, literature	Professionals
Technology	Primitive, R & D	Baling wire, but working	Tooling, rev control	Automation, China?
Tech People	MIT/ WPI types??	Teams	Labs, R&D mgt	Vice Presidents
Admin	none	Office manager	Formal Organization	Vice Presidents
Admin People	See above	Small staff (teams)	HR, Purchasing	Vice Presidents

Figure 9-1. Stages of growth

When I encounter early stage teams in my consulting work, I ask that the visible managers complete a "clean" copy of the sheet shown in Figure 9-1 and try to identify where they are in the evolution and maturity of each function. That set of observations is usually quite fascinating as the teams are still quite small and usually offer more consistent views.

To illustrate this, let's look at the Accounting function. In the startup phase, simple Excel models executed by members of the team are sufficient. Later, packages such as Intuit's QuickBooks are required because of the number and complexity of transactions required to sustain the business. The team can anticipate this by training people, hiring consultants, utilizing online services, and so on. There is a particularly unsettling trend among outside investors to say to the startup team that they are not experienced enough to grow the company and either fire the team (or individual members), as allowed by the terms and conditions of the investment, or move them to less visible positions. Either outcome can be avoided. Beyond the excitement and organizational flexibility of a startup company, few want to stay in that fragile and unstable state. Individual growth becomes important and valued.

The Legal Structure

Early in the process of commercialization, certain organizational decisions need to be formulated. An important one is to understand which corporate format serves the project best. Generally, the nuances of this decision are complex enough that they remain in the space of legal opinion. Although the term is broad, a simple form of incorporation that creates companies is most common. But even incorporation has multiple paths. There may be differences in incorporation details on a state filing level, but they always ask for these:

- Business purpose. This includes a general statement and detailed definition.
- Corporate name, including identifiers such as corporation.
- Registered agent, including a hard physical address for that person.
- An incorporator. The person (usually an attorney) who actually files the Certificate of Incorporation.
- Share per value. Usually a fictitious value such as $1 per share.
- Number of authorized shares of stock at the time of filing. In Massachusetts this number, although arbitrary, sets the filing fee.

The concept of a corporation as we know it today has its roots in the Latin term "corpus" which means "body of people." Its significance became more prevalent in English law, where royal charters were granted to certain trading merchants to protect their business territories and allow the formation of taxable entities. According to the text *The Anatomy of Corporate Law*,[1] the real push to create the modern form of incorporation came with the Industrial Revolution of the mid-1800s whereby the commercial need for business structures outstripped the government's ability to issue royal charters.

We will soon see that there are multiple forms of modern incorporation. In common, they have certain attributes that serve the commercial project's journey well. To list a few:

- The corporation has a separate legal identity. It is separated from the owner's assets. This allows it to do commerce, write contracts, and accept funds independent of the ownership.
- Consistent with that, it can sell shares in public stock trading markets and participate in private equity offerings.
- It allows governances by an independent board and can delegate operating authority to management.
- It allows for an orderly liquidation of assets and equity. This allows individuals to continue employment through changes of ownership.
- It can participate in certain tax incentives that are not available to individuals.

Within the term "incorporated," there are multiple nuances worthy of consideration. They are embodied in tax, government, and shareholder reporting, and scope of ownership attributes. Some of the common ones include LLC, proprietorships, and partnerships. Each has advantages and limitations. An LLC, for example, avoids the double taxation on corporate profits and individual dividends to its owners. It limits the number of shareholders to 75, which might limit the organization's ability to raise capital. The LLC model is not available to banks and investment banking organizations.

In addition, there are tax considerations offered in Chapter C and Chapter S filings, which focus on the tax obligations of stock earning distributions and equity participation/obligation rights. It would be tempting to try to delineate the differences here, but the exact combinations of tax, legal, and ownership attributes lie in the hands of the legal and accounting professionals. The commercialization/market dynamics aspects would be an overlay to their recommendations.

---

[1]Reinier Kraakman, John Armour, Paul Davies Krashum, et al., Oxford University Press, 2004.

The Mechanics

Beyond the legal and tax considerations are the tactical aspects of how the organization works. How decisions are made, how is capital distributed, how do reporting functions operate, and how do budgets/schedules work? These options are best described in a series of hierarchal models envisioned as functional, product/technology, and market driven. Let's look at each in the following sections.

The Functional Option

The most prevalent form of organizational structure is called the functional model. It is a hierarchal structure that starts with direction from the board of directors and operates through the Chief Executive Officer. It relies on a reporting structure of functional reports such as the Chief Operating Officer (COO), Chief Financial Officer (CFO), Chief Marketing Officer (CMO), and so on. Because of these titles, the level of management is called the "C" level. In other variations, these folks are referred to as Vice Presidents and the C nomenclature is not used. IBM and many other large organizations have promulgated a "rule of six" that allows only six reports in each level of the organization. An example of a functional organization chart is shown in Figure 9-2.

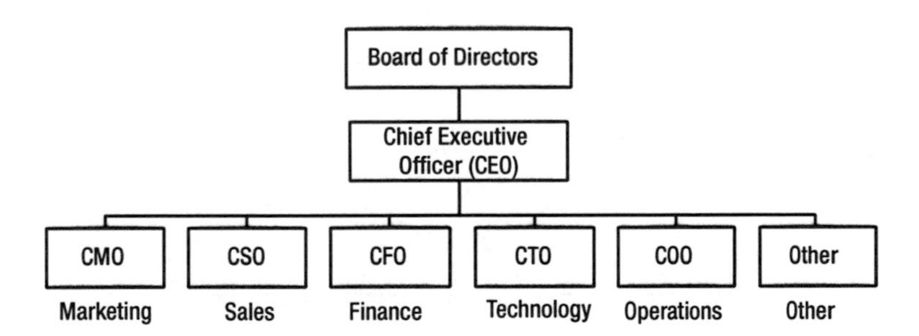

Figure 9-2. Sample functional organizational chart

Although this is a generic version of an organization chart, it reflects the rule of six, which suggests that organizational models of reporting that optimize around six reports. The "Other" area reflects that these models must be adapted to the needs of each organization. For example, if they were a pharmaceutics medical device project, it would show a C-level regulatory position. Similarly, if it were a multinational organization, it might have a C-level international role.

The advantages of this model are that they allow the depth of specialization to work. Skillsets and experience can be combined to become more effective and thus results in more efficient operations. Productivity can be measured and is usually considered higher in this approach. I suspect that is why it is so popular. All of this is especially true in earlier stage organizations where the "bench strength" tends to be limited. It is not without its drawbacks as it requires top management skill in ensuring the clarity of goals and effort to keep the functional areas separated. Coordination is an additional responsibility to avoid the risk of the "silo" mentality, where the groups do not coordinate with each other. Both of these risk factors can be managed.

The chart shown in Figure 9-2 implies a finished and completely mature organization. This is rarely the case. Personnel changes, hiring challenges, budgets, and most importantly stages of growth enter the dialogue. Early stage organizations thrive on informality and thus lack of structure. With growth there is an increasing need for more formal approaches. There are also other models, as discussed next.

Product-Based Organizations

There are many alternative forms of organizational structures. A common variation is the product-based divisional model. This choice is usually driven by customer grouping or product attributes. A classic example is the General Motors model, whereby there is a Chevrolet Division, Pontiac Division, Cadillac Division, and so on. Each division carries its own financials, marketing, sales, and manufacturing capability. A clear advantage of this approach is the ability to create strong and differentiated product identities. One wonders what the efficiency and productivity issues might be as individual functional roles are developed. A sample model might look like the one shown in Figure 9-3.

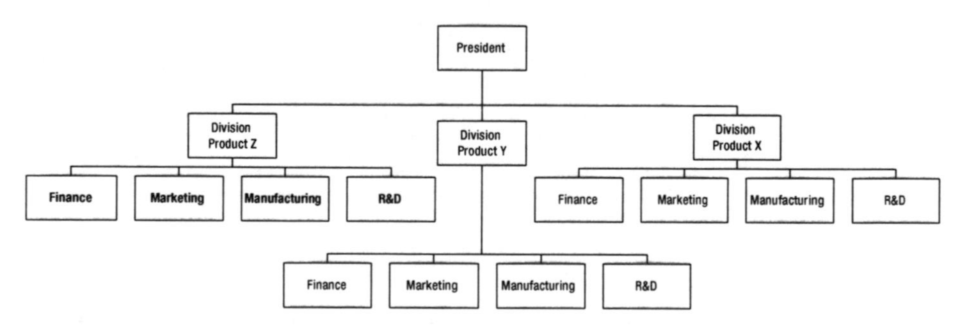

Figure 9-3. Product-based organizational model

One might argue that all cars have tires, radios, engines, and so on. Yet the marketing folks might push back on the need for differentiated product lines to meet the needs of different customer demographics. There is certainly the space for additional variations on these two approaches.

Other Approaches

Organizational structure design is more of an art than a science. The organizational formation may be ruled by strong personalities, industry mores, or market drivers. Each carries its own attributes. Each is controlled by the company's maturity or the project's growth. Matrix organizations whereby divisions are made cross-functionally were in vogue in recent years. Today there is a push toward flatter organizations, where much of the hierarchal and bureaucratic influences are diminished. The needs of the project best control the decision process. Commercialization success should be the decision arbitrar for the creations and change of the model. Organizational change is always disruptive, yet the model is always under pressure to change from customer and market needs and of course from growth.

One alternative approach was employed by IBM in the development of the Blade server platform in 2006. In a paper presented to the *Journal of Product Innovation*,[2] Charles Snow, Oystein Fjelderstad, and others wrote "Organizing Continuous Product Development and Commercialization: The Collaborative Community of Firms" and traced the development of the Blade server product system and how multiple firms were invited by IBM early in the project. In doing this, IBM acknowledged a) that they would need help in bringing the technology to market and b) it would bring the resultant project to market faster. The authors also noted that the market projection for the technology was somewhat uncertain and that this model was built in other organizations that were committed to its success and acceptance. Snow et al. also acknowledged the potential for conflicts and communication problems, but felt that the presence of a strong central source like IBM would anchor this.

Moving Forward

Although a great deal of time and effort is allocated to the choice of the specific organizational model to be implemented for a particular project or startup team, there are several themes that may control the discussion. They include the following.

The place in the lifecycle of the organization. In early stage entities there is a given sense of informality. It is a team in which job descriptions are fluid, distribution of tasks is flexible, and time domain pressures and deadlines are met by the whole team. Technical competence is rewarded (and valued). Governance and performance metrics are secondary. Compensation is not formalized and usually less than market equivalent as the organizations strive to preserve precious capital. Individual characteristics and contributions

[2]*Journal of Product Innovation Management* 28 (2011): 3–16.

of the players become important attributes at this stage. Some individuals thrive in this space and become ill-suited for the demands of the maturing organizations.

As the projects move forward, the number of personalities increase and the beginnings of the growth model appear. Functional teams emerge, informal job descriptions are utilized, and the beginnings of cohesive compensation procedures fall into place. In addition to the technical competence so critical in the early stages, leadership and communication skills increase. External resources such as consultants and other service providers become part of the team. Not all of the early stage individuals are suited for this path but infusions of training, consultants, and mentor-based coaching can help.

A third stage of maturity enters with the appearance of the professional disciplines of Human Resource professionals, job descriptions, and more formal performance reviews. Sometimes external pressures such as financial reporting and regulatory needs such as compliance reporting set the timing of those changes. Certainly customer needs remain a high priority and functions such as customer order fulfillment and after-sale support enter the mix.

Finding, Motivating, and Keeping Talent

As the number of individuals increases, the formality of their interactions occurs. With that, the need for acquiring these resources increases in scope and complexity. Professional recruiters (head hunters), job fairs, and employee referral programs now play an important role. The challenge of finding talent changes with the development of the project. Early stage entities rely on networks and informal connections. The needs and processes to fulfill them becomes more structured as they progress.

No matter how formal or structured the process is, a useful stating point is to articulate a job description. This document is important in that it:

- Defines the basic job scope and responsibilities. It becomes an implied contract for tasks to be accomplished and its deliverables. It also defines boundaries that allow others to see how they can interact.
- Becomes a measuring point for reviews and performance measurements.
- Delineates reporting responsibilities, both upward and downward.
- Assists others such as recruiters and outside personnel to help find and secure needed talent.
- Is pliable enough to be reviewed and changed as the projects progress.

- Can be employed throughout the organization from the board members to hourly workers.
- Defines the experience and skillsets needed to succeed. Sets up training metrics.
- Allows comparison to industry trends (surveys) and other comparable metrics.
- Becomes a "straw man" in early stage projects to capture position information of unfilled slots. This is particularly useful in writing and executing business plans where the future needs are articulated, but not filled.

There is very little, if any, downside to the discipline of job descriptions.

Acquiring talent is one side of the equation; motivating individuals to perform in concert to the needs of the project or organization becomes a major threshold for success. Much has been written about teams and teamwork. There is a quote credited to Helen Keller (citation unknown) that is a favorite of mine. It reads, "Alone we can do little; together we can do so much."

Some of the elements of teams that apply to commercialization are:

- They offer an opportunity for collaboration that can bring the synergy of different disciplines to bear on the outcome of the project. This allows progress without the burden of using parent organizations.
- They allow change and transitions to occur because the various collaborative players own a "stake" in the outcome.
- They can incorporate unique performance metrics adapted for the specific tasks.
- Although they offer a new chance for conflicts that must be managed, they also allow potential barriers to be disrupted earlier.
- They offer new opportunities for leadership skills development among the participants.
- The management details are smaller than the parent organization's, which offers a granularity of performance management that can flag issues earlier.
- New channels of communication between existing groups can be developed in teams.
- They allow individuals to assert responsibilities not offered in the existing organization.

Balanced Metrics

It's not easy to sketch out various organizational schemes. You must define how the operation's tasks are to be integrated into the parent organization. This isn't just operational issues of space and the organizational capacity to manage them (Human Resources, Maintenance, and Benefits), but also addresses the issues of integrating them into the parent organization.

One of the most prevalent measuring tools comes from the use of operational budgets. They are derived from the overall planning effort cited in earlier chapters. Here the flow of expenses and capital investments is usually presented in monthly segments. This flow delineates personnel, equipment, services, and even outside resources such as design firms and consultants. If it's done well, the categories of the budget align to the parent organization's accounting system so that they can be integrated into the overall corporate metrics.

Can innovation and creative processes be measured? The simple answer is "of course" in that indirect metrics of milestones, productivity, goals, and expenditure can be monitored and analyzed. Although they are indirect, they offer a measurement tool for perceived progress. Without at least this level, you're flirting with chaos and financial overrun. Metrics also allow for the comparison of one project's approach to others. Simple percentage comparisons are usually sufficient for monitoring progress. This type of evaluation also allows for contingencies when unplanned roadblocks occur. Budget allocations can show how resources can be shifted to meet these unexpected needs.

Creating the Environment

Organizational dynamics certainly embrace structure, finance, metrics, and the art of finding, motivating and retaining talent. From outside the team, there is a much more subtle and controlling force in how the groups operate. Whether in large or startup entities the couture that controls the workplace and eventually embraces the customer is paramount. Beyond the excitement of starting a project, there remains a major challenge of sustaining an innovative or entrepreneurial mindset in the day-to-day activities.

There are many tactical elements to sustaining that style. Open meetings, sharing corporate progress and issues, rewarding risk taking, and encouraging innovative players are just a few. Even Friday wine parties (Silicon Valley) are known to work. Company benefits and financial incentives like part-time tuition fall into this category. Strong leadership styles and employee-centric working models also fall into this area. I suspect that most of the successful models are a mixture of tactics and art.

Summing Up

The choice and implementation of the organizational team model clearly affect the outcome of a project or a new venture. Like the artist's work, it is constantly under scrutiny and receives endless tinkering. Attention to the performance goals, action plan, and measuring metrics of progress are the livelihood of the managers of this activity. Each element contributes to the overall success.

Success can certainly be measured. In the next chapter, you look at how the financial decision process operates. Large organizational models and startups are compared. In addition, the role of outside capital decisions is examined, both from early stage venture backed projects to the realm of investment banking and public market decision making. Finally, the role of non-dilutive funding such as government and foundation grants is put into perspective. That will allow us to see their impact on the probability of success in technology commercialization projects.

CHAPTER

10

ROI: Does It Make Sense?

The opportunity is present, the plan is in place, you have access to capital, and you can assemble the right team. Should you go ahead full tilt? You need to step back and evaluate the overall project. Does going forward make sense in the context of corporate goals, IRR of cash flow, the organizational balance sheet, and the cost of capital? Eventually all this must be rationalized in the go/no-go decision. In many cases, this also engages the BOD and the governance model of the organization and project decision processes. This chapter will give you a method for deciding whether to pull the trigger—or not.

We now know that there are multiple influences in the decision processes surrounding technology commercialization. The importance of market-driven needs, creativity and innovation, solid organizational models that are nimble and adaptive, and the planning details and performance monitoring are important for sure. Capital plays a critical role as the projects will need it to go forward. It is the fuel that drives the engine. It can be metered out in lean fashion or significant resources can be brought to bear.

There are multiple sources of capital that range from internal corporate funding to formal investment capital led by angel and venture capital groups to public funding. Each has its own nuances in their priorities, decision-making processes and, of course, influence on the outcomes. Let's first look at some of the alternatives and their investment goals. They fall into three broad categories.

Early Stage Capital

Particular to the formation and early stage standalone companies are three subcategories. The first is sometimes referred to as FFF (Friends, Family, and Fools). The Fools aspect deals with investment made well before the project is considered an investible entity. These investments are made on an intuitive or emotional basis rather than one subjected to critical and professional review. Being early in the investment cycle it also requires enormous returns to compensate for the time of the investments. A variation of this are projects that are self-funded or "boot strapped." This type of entity has the freedom of not being subjected to the onerous Terms and Conditions of more formal rounds. If a project is capital-intensive, this form of investment becomes limited by how "deep the pockets" of the entrepreneur are. The structure of these deals is also more informal and may even be fulfilled by simple partnerships.

Angel Investing

The next category of early stage capital infusion deals with more structured investments characterized by angel and venture capital. Angel capital as a source was identified almost 40 years ago by Professor Bill Wetzel at the University of New Hampshire. Bill was interested in how wealthy individuals living in New Hampshire were investing in early stage ventures as individuals. He noted how much more effective they would be if they worked as groups. Many of the individuals were quite independent and not interested in this. Today this category of investors have grown and have a national "trade" association called the Angel Capital Association (ACA), which is comprised of 187 chapters across the United States. There are 32 in New England alone.[1] They meet in national and regional sessions to discuss trends, best practices, and even have managed to consolidate deals into projects. Angel investors invest their own funds (even if grouped on separate LLC arraignments). Returns are measured on individual portfolio base. Their motivations to invest are broad and include the need to "give back," as well as harder investment metrics.

The mechanics of an angel meeting is interesting to note as it gives us insight into the investment-decision process. Angel investing has emerged from an individualistic single-person decision. As the potential project stream increased from the early Wetzel observations, more disciplined approaches emerged. Membership in angel groups is varied but sometimes the groups form a "personality" about which type of deals they favor in terms of sector, size of investment, and stage of growth. An example is Common Angels in Boston. Their focus is on "software, Internet, digital media, and cloud, from Seed to Series A."[2] Disregarding

---

[1] `www.angelcapitalassociation.org`.
[2] `http://commonangels.com/`.

this opening preamble wastes time and effort and sets up a rejection that is not based on the merits of the project. A rejection is a serious issue as it propagates thought the investor community.

Active angel groups have memberships numbering 50 to 75. This number allows independent internal committee structures to evolve. An example is the screening committee whose role is to pare down approximately 20 projects a group receives in one month to the one or two they will act on in an open presentation. The committee screen focuses on their merits as a potential investment and the fit to the group's interests. A strong priority is shown to referrals. If a member or known service providers have been identified as the source of the lead, it has screening priority. Most groups have provided Internet-level information for the first encounter. Beyond the political implications of this, I suspect that it also validates the alignment of the deal to the group's values. This duality is shown in Figure 10-1.

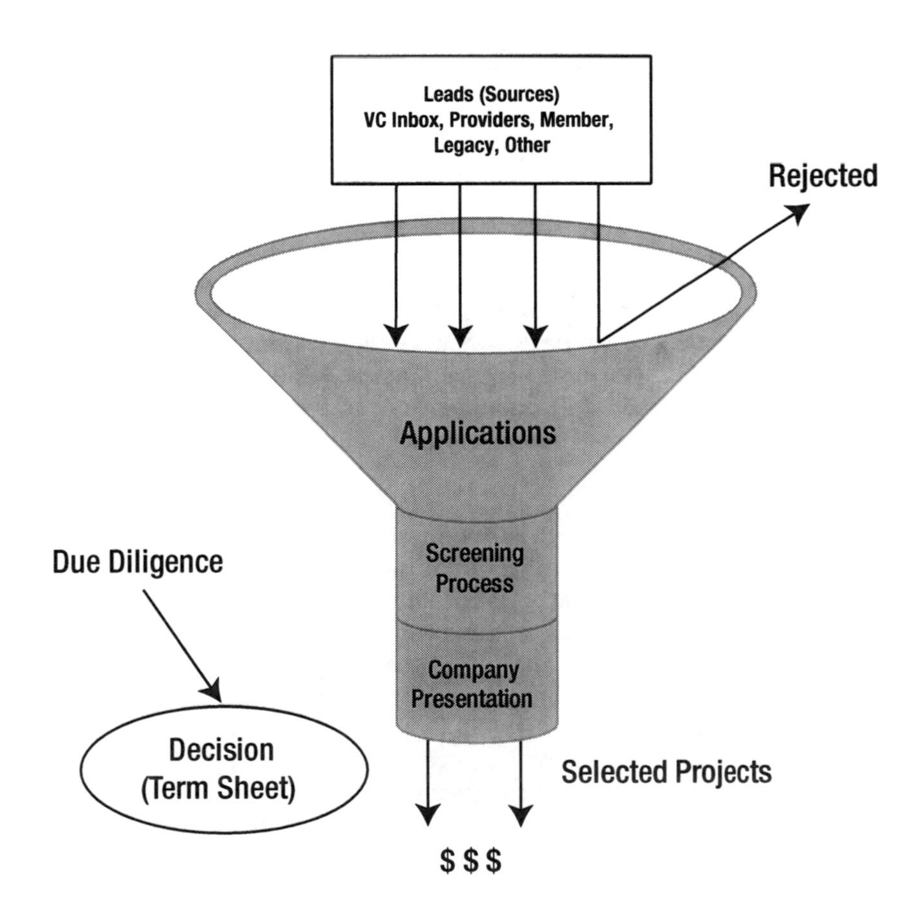

Figure 10-1. Traditional angel funnel

As a given project negotiates successfully through the screening process, it is presented to the general membership in a somewhat formal (usually organized at PowerPoint level) presentation. Group meetings are typically scheduled on a monthly basis. The presentation is characterized by an individual investor stating his or her investment interests. "I'm in" is a typical phrase. An awkward silence is an alternative. Those of you who have seen the *Shark Tank* show on ABC may recognize the sentiment. If a sufficient number of individuals buy in (remember it is an individual decision in angel investing), the project moves to a due diligence committee process.

"Due diligence" is the process of investigating the material facts and what attorneys call the Representations and Warranties of a given project. The process has its roots in the U.S. Securities Act of 1933, whereby stock brokers were required to check the material information with respect to selling stock. It was particularly aimed at potential investors who may have lacked sufficient information about the risk elements of a particular investment. There is an international equivalent in the Foreign Corrupt Practice Act (FCPA). With the trend toward increased global commerce, this provision becomes even more relevant.

Upon completion of the investigation, a formal (usually written) report, complete with recommendations, is submitted to the general membership. An important element of the report is a proposed outline for the Term Sheet that accompanies the group's proposal to the presenter. This document sets the basic parameters of the offering. It is a formality of a succession of documents, including the Letters of Intent (LOI) and Memoranda of Understanding (MOU). They set the intention of the deal with an outline and future commitment to settle the details. The National Venture Capital Association (NVCA) maintains a catalog of common deal document templates on their web page (see `www.NVCA.org`), including common Equity Term Sheet elements. They include:

- Amount to be raised and the timing of the closing transaction.
- Price per share, which is the intersection of money and equity. In many ways, this is the heart of perceived value of the project in both current and long-term (3–5 years) timelines. Sometimes it is expressed in a percentage of the total equity (ownership) of the company. It is a swap of common shares for money. There is significant misunderstanding about owning 50% or less and control of the company. If an investor issue relies on the financial positioning of the players, it suggests that there are much more significant issues. Later in the Term Sheet, more onerous control and reporting issues are defined. The argument of whether you would want to own 10% of a sufficiently well-funded and ongoing enterprise or 90% of floundering one is quite obvious. Recognized value and the ensuing

wealth creation depend on the number of shares owned as well as their perceived value. If all that weren't complex enough, there are various classes of stock ownership such as "preferred shares," whereby the owners have both liquidation preferences and preferred voting positions.

- Among the preferences enjoyed by the investors as preferred shareholders is one called the "Liquidation Preference." This is a diabolical tool invented to ensure that investors can "cash out" of a deal first. It goes something like this. If a company raised $10 million of venture money for 30% of the company's equity and the company is sold for $25 million, the investors should realize $7.5 million if distributed on a prorated basis. Under the liquidation preference provision, they would receive $10 million. This is called a 1X liquidating preference. If the multiplier were 2.5, they would receive the full $25 million—yikes! In the March 8, 2014 issue of *Business Insider,* author Nicholas Carlson called this a "horror show." I think his characterization is an understatement, but the concept of liquidation preference seems non-negotiable in the world of investment capital. Regarding an investor as a partner in the deal also seems a stretch.
- There are other provisions of a Term Sheet such as anti-dilution formulas that protect the investors in the likely case where additional stock has to be issued to raise subsequent rounds of capital. There are voting rights in decisions regarding selling the company and even raising of additional capital. These are generally referred to "Registration Rights," whereby under the Rule 144 of the Securities act of 1933, investor shares can be "registered" in a manner that allows them to be sold publicly. In effect, it allows the investors to sell the company at their preference—yikes again! In addition there are participation rights that give them voting seats on the board of directors and rights to terminate the CEO's employment.
- The NVCA template cited previously suggests that Term Sheets are somewhat "boilerplate" in their content and format. This is probably somewhat true. But it also suggests that the crafting of the terms is clearly in the domain of corporate attorneys. It also suggests that the terms are somewhat lopsided in their definition in favor of the investors. Solid management performance and the meeting of goals and schedules make it an act of absurdity for the investors to capriciously act to invoke the onerous provisions.

With the Term Sheet proposal secured, an offer is assembled for the presenter. A period of negotiation about the terms and price per share follows. In addition, the details and plan of action to go forward are ironed out. Another interesting aspect is that angel investors must declare that they are accredited investors. Under the Securities and Exchange Commission (SEC) Regulation D, an accredited investor is one who has an income of $200,000 and a net worth exceeding $1,000,000. Its intent is to ensure that the investor is both sophisticated enough and substantial enough to make high-risk investments. There are initiatives such as the Dodd-Frank Act of 2010 that try to exclude the primary residences in the $1,000,000 calculation and Government Accounting Office (GAO) attempt to raise the require to $2.5 million and the current income to $300,000. These proposals would help limit unqualified risk, but also limit the number of potential investors.

One significant aspect of this step is the selection of a lead designated board member. This person not only sits on the board, but also carries the voice of the investor group. This aspect of the investor membership also carries the responsibility of reporting back to the investor group. The formality of the process varies with size and with the history of the investment portfolio. To further amplify this process, it's fruitful to compare the angel investing process to the more formal role of the venture capitalist. To gain a perspective of this comparison, see Figure 10-2.

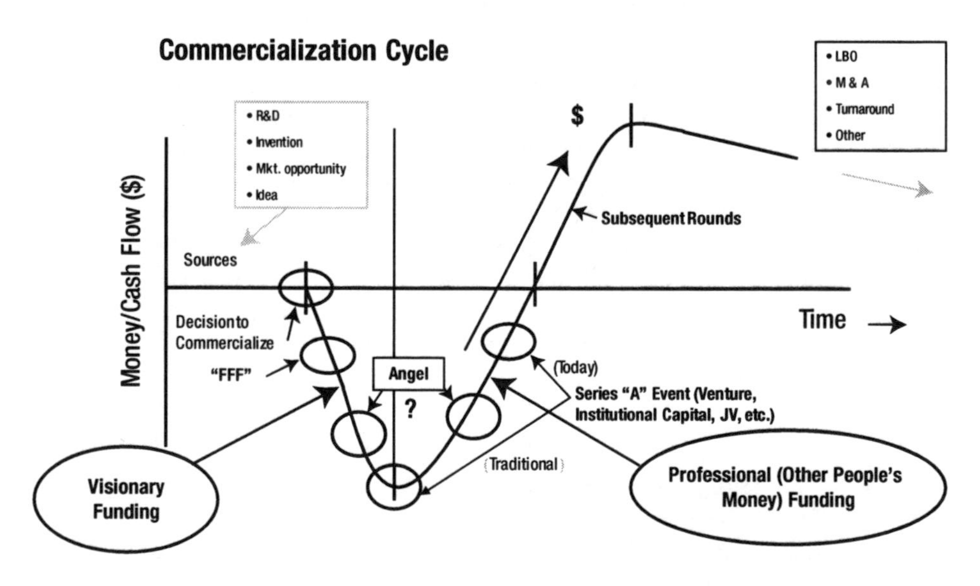

Figure 10-2. Commercialization cycle

Institutional Investing, the Venture Business

In 1945, Georges Doriot (later coined the "father of venture capital") founded the first venture capital firm in the United States in Boston. It was called American Research and Development) in a text entitled "Creative Capital: Georges Doriot and the Birth of Venture Capital"[3] and was co-founded by Karl Compton (former president of MIT) and Ralph Flanders, a businessman. What distinguished his investments was that they were made with other people's (limited partners) money under his control. This idea was validated by an investment of $75,000 he made in Digital Equipment Corporation in 1957. The stock was liquidated by an initial public offering in 1968 at $355 million. That was an IRR of 101%. From these beginnings, a new industry of venture capital with centers in Silicon Valley California and Boston was spawned. Doriot was born in Paris and came to America to attend the Harvard Business School, where he later became a professor. He was also known for his pithy quotes. One of my favorites is, "Without actions, the world would just be an idea." He also preached that the expanded role of investments should be one of "nurturing" the projects beyond just the financial mechanics. Unfortunately, this secondary role has shifted from focus in the modern incarnations of venture capital industry. It has become transaction-oriented.

Comparisons

Within the categories of angel and venture capital funding are the primary sources of capital for early stage, standalone, ventures. Although they both focus on early stage projects, the differences are significant. An obvious starting point is that the sources of capital are quite different. Angel funding is private money belonging to the investors. Typically it derives from cash out of a previous venture and controlled by the individuals making the investment. This is significant because it allows a much broader agenda for the decision process. I was the co-founder of the Cherrystone Angels in Providence, Rhode Island. In that fund we became so interested in the "other" aspects of the decisions that we actually surveyed the membership at each investment as to why they made the specific investment beyond ROI considerations.

The results were intriguing and ranged from some who "wanted to give back" to helping economic growth in Rhode Island. Sometimes this secondary "agenda" is useful to motivate individual investors to serve as mentors and even get their possible board participation in the new ventures. Sometimes their participation is more tactical and includes industry contacts and customer introductions. The formal investments/equity interactions between

[3]Spencer E. Ante, Harvard Business School Press, 2008.

angel and venture investors are somewhat similar, as they both employ similar Term Sheets as part of their investment packages.

There is a minority view that the angel groups benefit from maintaining "side car" funds managed by the leadership of the group. In theory, this structure allows subsequent funding to be applied that is independent of the fluctuations of individual investment decision processes and is closer to the funding need dynamics of the project.

There is now a brokerage of sidecar funds established by SideCarAngels.com. It was founded by Rick Lucash of Launchpad Angels and Jeff Stoler. Common Angels, Houston Angels, and Golden Seeds Angels are cited by the Kauffman Foundation-funded Angel Resource Institute as examples of angel groups that have these funds. The management of the funds is a different dynamic than individual decisions and closer to the managed funds of venture capital. The hybrid is touted as being effective but there is not sufficient data to support their overall effectiveness in the investment cycle. Anecdotal data from the Angel Resource Institute[4] suggests that about 20% of all angel groups have these sidecar structures.

Venture capital is more structured than angel involvement and tends to appear later in the lifecycle of new ventures. "Funds" within venture groups are created based on proposals to groups of limited partners. Pension funds that require small parts of their portfolio to be invested in high-risk/high-growth help boost their overall portfolio performance. The Employment Retirement Income Security Act of 1974 (ERISA) set the stage for pension fund involvement. The California Public Employees Retirement System is a good example. The 1970s also saw the inception of some of the larger firms that became thought leaders in the industry. Kleiner Perkins in Silicon Valley, California and Greylock in Boston are examples of these firms. In addition the National Venture Capital Association (NVCA) was formed in 1974. It became the ultimate source of information and dissemination of information about best practices for participating firms.

Venture funding became a "growth industry" for almost 40 years until March 2000, which is when the infamous "dot com" collapse occurred. Too many deals were focused on soft metrics such as "eye contact" and "stickiness." Attention to longer-term value propositions slipped from focus. The investment deals faltered and the result was a major meltdown of funds' values. The impact of this was that going forward, venture fund attention and investment priority was given to less risky and more mature deals.

---

[4]www.angelresourceinstitute.org.

An example is the computer disk drive industry. In the 1970s, there were 35 domestic manufacturing firms funded by the venture capital industry.[5] Four of them could supply the world's need for product. Today, there are four major suppliers. The others failed, merged, or simply fell by the wayside. Clayton Christensen, cited in Chapter 1, had a more severe view. It was that the industry, fueled by venture capital, simply couldn't adapt to technological change fast enough. Whatever reasons are assigned, it is clear that there was simply too much money chasing bad deals at that time.

In the lifecycle model, there is a blurring between the role of the venture industry and the role of investment banking. The venture firms retreated from early stage deals as a reaction to the failures of the "dot-com" companies and left behind the funding needs of the early stage deals.

If the angel industry had not grown to replace early stage funding, one ponders about where the funds for new, high-growth potential projects would come from. The dramatic impact of the "dot com" portfolio model is shown in Figure 10-3. It is a graphic of NASDAQ Composite Index, which is technology stock laden. It presents the rapid gain and loss of the value of their stocks. Of more significance is the strategic change in venture funding that occurred. It favored later-stage deals that embraced less risk.

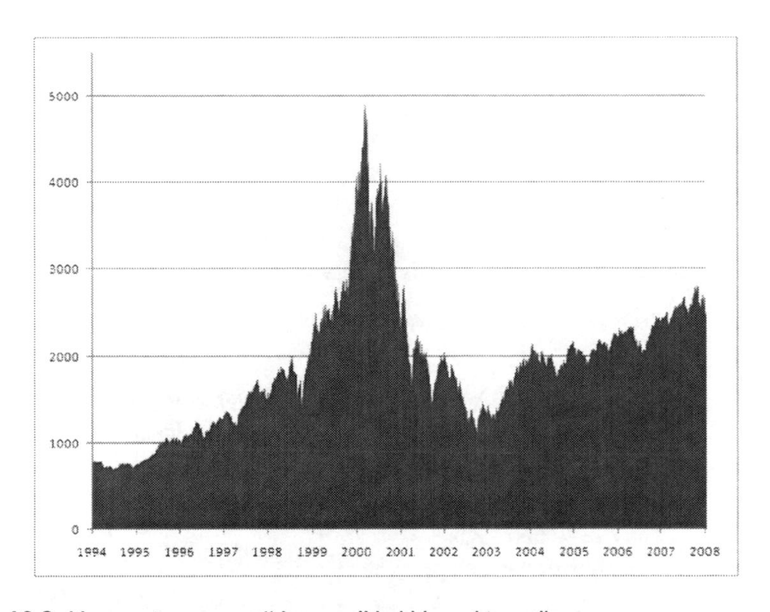

Figure 10-3. Venture investment "dot com" bubble and its collapse

[5]Disk Trend Report, James Porter, 1974.

With its 1.6 million person membership, 1-2% of its assets invested in this category are a small percentage, but a significant amount of funds. Typically there are multiple funds under management at any one time. The implications of the multiple fund models are that there are liquidations requirements of 5-7 years of the individual funds created that drive the timing of the liquidation events. Probably more significant is that the funds are managed by professional teams rather than the previously defined individual angel decisions. The team members tend to have sector area experience and may have even been in executive leadership positions in a given industry.

The focus is on the ROI components as they are the basis for the limited partner participation. Fee structures for these transactions are so insidious that individual partners are rewarded with a "carried interest fee" of several percent of the invested capital they place in deals. This fee has certain limitations in that the original amount must be returned and certain return "hurtle" rates must be maintained. The term of "carried interest" is derived from 16th Century shipping, whereby ships captains were given 20% of the profits for carrying certain products. Today it is meant to be an incentive to allow risk-based investments decisions. They are indeed an additional financial burden on the outcome of the investment.

There are clearly differences in those primary sources of investment capital. Today, each has its place in the lifecycle of the company or project. In the earlier days of the 1960s both seem to be converged on the startup. Venture capital went through a strategic shift after the much touted "dot come" bubble burst and there were tremendous loses and pullback. Today, venture capital that utilizes other people's money seems more risk adverse but is capable of more money per deal. Angel decisions are individual and can appear earlier in the cycle. What is important is that the relative positions and decision metrics are understood. Another view of this is shown in Figure 10-4.

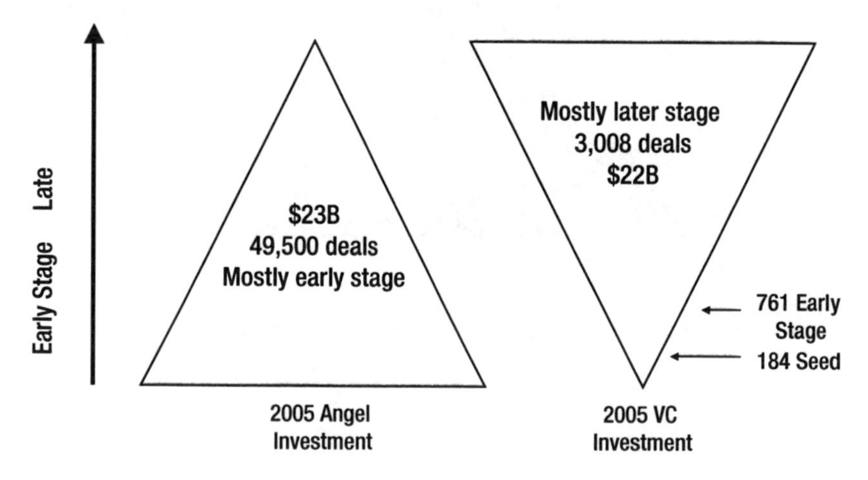

Figure 10-4. Convergence of venture and angel deals

The vetting process of the venture firms is not that different from the angel process and resembles a funnel model. A significant difference is the presence of a formal investment committee. Usually attended by general and senior partners in the firm, they are the equivalent of a general meeting of the angel firms. They have a charter of ensuring alignment of the funds' investments to the agreement secured with the limited partners who have supplied the funds upon which they operate. The difference is that this decision revolves about "other people's" money and the angel decisions are made on a personal (individual) level. That aspect separates the venture capital decision process from the intuitive aspect of angel decisions.

It is important to note that the angel deals tend to be earlier in the lifecycle, thus they require more subjective decisions as there simply isn't enough factual and operating information to support the decision process. Today the venture folks have moved to more substantive deals that occur later in the cycle. The implications of this are that they must also envision strategy of an exit to public equity markets and/or merger activity. The terms structure actually envisions such an exit and sets the rules for that later decision process. All this implies an orderly process of investment vehicles. In reality there are also venture fund that are referred to as "boutique" and focus on earlier stage deals in specific and narrow segments. Certain industries such as the medical products and biotech require longer and larger investments. They depend on specific industry venture firms earlier.

The Broader Resources

According to the NVCA (`www.NVCA.org`), venture backed deals accounted for 11% of private sector jobs and 21% of the GDP. Formal early stage capital is an important segment of the formation of new companies and ventures. However, it is not the entire landscape. In a personal tally, I identified over 20 sources of funding that ranged from personal credit cards to public, non-dilutive grants and foundation support Let's look at two of the these large areas.

The first category is one of non-dilutive infusions of capital. The term drives from the fact that money is brought to the project or new venture without an accompanying transfer of capital. In that sense it is quite attractive. On the other side, there is usually a deliverable of written reports, procedural changes, or a working prototype model. There is enormous freedom of choice that accompanies equity-based funding; maybe not so in the world of grants and alternative funding. Something about the saying that "there is no free lunch" seems to apply.

Certain fields use government funds such as National Science Foundation (NSF) grants in bio tech and basic science and in for forms of R&D funding. Fundamental research in medical and biological fields is so intense that an aggregated source such as the government can make that level of investment at the basic science level. Large and bold investments are part of certain industries. Yearly automotive changes, for example, require large annual tooling and technology expenditures. That is simply part of the automotive business model.

Certain government grants are particularly focused on commercialization. An example is the use of Small Business Innovation Research (SBIR) grants. Specifically focused on commercializing innovation, the SBIR was created in 1982 to help various government agencies secure the technology they needed. The U.S. Department of Defense is the largest grantor and issues over $1 billion annually. The grants are released in phases. Phase I grants are generally $150,000 for six months. Later phases may be over a million dollars. There are variations of SBIR grants called the Small Business Technology Transfer Program (STTR). It requires a partnership between a research institution (minimum of 30% participation) and a firm capable of taking the technology to market.

Philanthropic funding is certainly an important aspect of non-dilutive funding. An example might be The Jimmy Fund, which issues research grants through the Dana-Farber Cancer Institute. The science might be developed to find cures. License to industrial partners help the technology find paths to commercial reality.

One attribute of external funding is that it invites collaboration. Many academic or research groups combine resources and apply for grants and share their findings. Nonexclusive licensing models allow specific fields of use to be commercialized by individual applications.

Beyond infusions of capital, there are tax credits, incentives for loan energy offsets, and so on, which allow conservation of operating cash, but cannot necessarily be used for direct growth.

There are technical options in non-funding. An example is the use of "green shoe" issues of more shares than the company has issued. It is an overallotment of shares that can be used before the issuance of public stock (IPO). Certain combinations of debt/equity become attractive, such "mezzanine funding," whereby a lower form of debt is used that is later converted to equity. Vehicles like this serve to bridge financing rounds and offer the debt option.

There is another broad category of funding options. They might be referred to as "other." I once did an informal survey of this category and found almost 20 such sources. Included are the use of personal credit cards, mortgages on personal property such as real estate, informal loans or pre-payments from vendors and customers, and business plan competition prizes. The conundrum of this category is that the sources are generally not sufficient to meet the long-term requirements of the project or new ventures. Since time to acquire financial resources is such a precious resource, refocusing to primary, more robust sources is the priority.

With this category is the domain of parent company treasuries. This category is quite internally competitive in terms of demand on the resources. Mandates for ROI and IRR are delivered via internal committee documents and memos. The internal process for utilizing internal resources is controlled by hierarchal models that filter all the way to the board of directors for material or substantial amounts. Significant alternatives for acquiring funds in terms of absorbing debt and even floating stock to achieve the goal are available. It becomes the treasurer's ultimate "make/buy" decision. Sometime this resource can benefit from funding large capital expenditures such as plant and equipment or even corporate M&A activity such as other entity acquisitions.

At the other end of the spectrum, early stage companies do not possesses the "deep pockets" capability of supporting major capital projects of even those capable of high probabilities of success. In this case, the stage is set for raising capital by more conventional means.

Putting It Together

The purpose of this chapter is not to provide a litany of all possible sources of capital for new or innovative projects. It is rather to accent various sources of capital and offer the reader a perspective on the motives and goals of each type. When an airplane flies en route and encounters fuel starvation, there is an awful silence that emanates from the engine. Pilots are trained to begin procedures that will mitigate the risks associated with running out of fuel. Not so with entrepreneurs, innovators, and the teams that support them. Most assets can be procured. This includes people, technology, and even markets. The end of cash signals the end. In this sense, the airplane/pilot training analogy simply doesn't hold.

Each source of capital carries its own unique investor decision dynamics, expectations, and responsibilities. Understanding their nuances increases the probability of success in a commercial venture.

Once capital is secured and the project or venture is funded and launched, we will look at a myriad of regulatory and governmental boundaries on the modern corporate model. They are historically unprecedented in terms of the complexity, invasiveness, and demand that they put on management and the enterprise's resources.

Know Your Market

The challenge of quantifying the attributes and dynamics of the environment affecting the customer's decision to purchase a product or service is quite important to the eventual success of the commercialization process. It guides the allocation of resources required to accomplish the project's goals. You need a preliminary market assessment and a plan of action to achieve it. Doing so is often fraught with uncertainty to the point where the process is either done in a perfunctory manner or skipped completely.

One of the first implications of the planning process is how the complete operational chain responsible for producing the product can be engaged. This includes the formal planning process, inventory decisions, procurement strategies, work force expansions/contractions, purchases of capital equipment, plant and promotional activities, and so on. The cash flow implications of resource allocation to these functional areas is significant. The direction for these efforts comes from the market-driven plans that are the defining center of the planning activity.

A significant attribute of market assessment deals with the segmentation dimensions. Like the General Motors example cited previously, the same type of products (automobiles and trucks) can be used to fulfill the needs of multiple demographic and socioeconomic market segment needs. Sometimes the modifications can simply be cosmetic, such as body style changes. Other times they can be more substantive such as drivetrain or accessory kits available within a product line.

It is certainly an easy attribute to measure. The information can be directed to the product design and marketing strategy areas. It also allows the project to seek out profitability and competitive forces and can be chosen to be stable enough to yield positive results. It is common to set up a matrix of the various attributes and use the matrix to define the best path to market by rank-ordering the projects.

An example of this comes from the high-performance semiconductor chip manufacturers. Tooling for a new generation of processors is both expensive and time consuming. Market conditions change faster than this design process allows. Chip designers embed families of features in the chip architecture. As external market and competitive changes occur, they simply enable parts of the product. A subtle benefit of this measurement is that it favors customer retention because it helps identify the characteristics that favor customer loyalty and continued use of the brand name.

Decision Metrics

Management has the ultimate responsibility to decide whether a project (or company) should continue on the path to commercialization. This applies to a startup and to a continuing project within an existing organization. To make a good decision requires the successful implantation of three activities. They include the following.

Goals

In sports there is a clear correlation of specific goals and objectives to winning. There is also a need for rule books (and even judges). In the commercialization context, the projects or companies that are driven by clear and succinct goals that have been well communicated in the organization have a significant advantage over those that don't. The clarity also improves the decision making about project selection, capital expenditures, and potential profitability. Without clarity of goals or vision, decisions may be made in conflict, or in a suboptimal manner. Sometimes these goals are expressed in financial metrics. Other times they are expressed in market penetration or brand loyalty as expressed in terms of customer loyalty and repeat sales.

Computer-driven data enables us to generate significant amounts of information about markets, finance, and operations. If there ever was a need to explore this, it is in the space of sales and marketing. These functions drive the models and the numerical information helps guide the resources needed to realize them. It would be in terms of advertising expenditures or numbers of sales personnel required to complete the goals. This, in turn, creates a new need for assimilating that data and learning how to make decisions that utilize the information. New tools have emerged that allow better analysis and benchmarking against goals.

A common tool is the "Balanced Scorecard," created by Dr. Robert Kaplan of the Harvard Business School and David Norton, a WPI-trained management consultant. In their 1996 *Harvard Business Review* article entitled the "Balanced Scorecard: Translating Strategy into Action," they stated:

"The Balanced Scorecard retains traditional financial measures. They reflect the story of past events. In an era when investments in long-term capital expenditures and customer relationships were not critical for success. These measures are inadequate for guiding and evaluating information age companies to create future value through investment in customers, suppliers, employees, processes, technology, and innovation."

The Scorecard utilizes four perspectives that include Learning, Business Practices, Customer Perspective, and Financial Perspective. Of importance is that it looks for changes that delineate learning and growth. These attributes can directly be applied to the marketing function in terms of realized customer satisfaction and market penetration.

Reaching for Decisions

We live in world that has the potential to generate more data and information than ever before. With a mere click of a button we can access sources of information than could never have been imagined in earlier decision formats. Powerful search engines (such as Google) and new data-mining techniques have opened new and possible higher quality data than ever before. In the modern world of global competition and diminishing resources, the search for information reaches all corners of the globe.

With this abundance of data comes a need for a new level of sophistication in our ability to both interpret the data and utilize its information to make decisions. In addition, the rate of change is accelerating, as noted in shorter product lifecycles and nimble market shifts fueled by the Internet and handheld technology. In all of this, first principles as a basis for decision making are even more important than before. Let us look at some of the fundamentals:

- Adherence to the controlling vision. Possibly the most critical element of sound decision making is the comparison to the overall themes or vision of the enterprise in either corporate or project level decisions. It is enhanced in large capital investment issues, where the implications can be multiyear and beyond the lifecycle dynamics of the initial decisions. In product sales, for example, the aftersales margin contribution far exceeds the initial sales transactions. In addition, vision is not a static declaration. It is a dynamic resolution of changing times and markets. The ultimate responsibility to manage the vitality of the vision rests clearly in the dynamic exchange between the CEO and the board of directors.
- Monetary constraints. The "cost of capital" becomes an essential element of the project decision process. In the use of internal funds, there is constant competition. It is rationalized by the Internal Rate of Return (IRR) calculation, which allows the competition of returns between new investments in sustaining a current one. It represents a threshold of return that must be met or exceeded. External funding costs are controlled by interest rates for debt and the balance sheet impact of equity financing. Within those variables is the impact of sectors or market positrons. It requires an enormous amount of capital to create a new steel mill as compared to an Internet software enterprise. This sometimes forces a bias toward lower capital projects, with all the implications of that trend.

- Human resource considerations. All too common is the litany of cases that fail because of the team allocated to market success. The care of defining the required skillsets as articulated by robust job descriptions probably is larger than other considerations. Finding the right people, motivating them to succeed, and providing the proper resources to enable them to execute their responsibilities are critical.
- Technology/IP. Perhaps one of the more fluid aspects of a project is the status of technology. Most of all, is it mature enough to sustain commercialization? Has it been validated? Are there multiple sources of material and parts and can it be replicated? Is it unique and is it a "platform" that can be extended through multiple product lifecycles? Can it be protected with IP, and if so, how strongly? Correcting any of these issues after a product is launched can be expensive in terms of capital, time, and prestige. Consider the case of government-regulated automotive recalls as an example.

Market and External Competitive Forces

Clearly, decisions to move projects to commercialization don't take place in a vacuum. They are tempered and motivated by opportunities and by the changing dynamics in the marketplace. Changes in competitive offerings, changing regulatory climates, changing technology, and/or changing global influences are just a few of these forces. Traditional marketing tools such SWOT analysis, potential customer surveys, and general organizational awareness are the best mitigators of these changes.

Decisions and Intervention Based on Improving Overall Performance Against the Goals

Although there seems to be an endless list of influences and changing dynamic conditions surrounding the decision-making process, it is still inherent that a decision to go forward (or not) be made. Historically, statisticians such as Erich Lehmann offered theorems such as the Hodges-Lehmann estimator that try to find a statistical median. Likewise, data-finding tools and Bernoulli's St Petersburg Theorem try to find non-exceptional data to rationalize these decisions.

Inherent in this text is the premise that best results are realized by adherence to the central goals and/or mission of the organization. This allows coherent utilization of resources and their allocation and increased probability of success. In addition, it allows a dynamic feedback to the vision statement, which will allow it to be current with changing external decisions.

CHAPTER

11

Big Brother and Global Competition

Invasive for sure, the government's impact on commercialization is enormous. A long alphabet soup of interventions such as OSHA, EPA, FDA, IRS, SEC, PTO, UCC, CE, UL, Sarbanes-Oxley, tariffs, and other applicable laws and regulations affect all technological projects. Issues of control, reporting, and constraints hobble the very best commercial opportunities. Tomes have been written about the massive political and economic shifts taking place at a global level—in material resources, capital, human talent, manufacturing bases, government support, and more. While no one really understands the impact of the shift of manufacturing and services to Asia, for example, it's clear that there are new metrics for success. This chapter shows how people and companies compete in a world racing into the future.

We live in a time of unprecedented government and legal involvement in the way we start and operate modern enterprises. If this weren't complicated enough, in the United States, global pressures and competitive forces have increased the complexity of external involvement exponentially. Some argue it is better and some argue the other way. Whichever way the judgment call falls, we can agree that these trends can't be ignored. The government (both local and federal) significantly impacts the operations and responsibilities of a modern project or company. Their combined influence severely impacts its potential financial performance.

The application of various laws and regulations is not comparable in other countries. This leaves projects developed in a strong regulatory environment less able to compete. Compare manufacturing in the United States and China and you see China's imminent advantage of minimum environmental constraints, low wages, and nonexistent commercial rules that are so imbalanced that competitive manufacturing by U.S. entities is severely handicapped. This chapter examines those influences. By the end we will deduce whether they are good or not for improving the odds of commercial reality. In either case, we will see how they are integrated (intertwined?) into the total commercialization models.

A Perspective

A strange observation about governmental regulation processes is that governments and businesses have common goals of promoting prosperity and growth while protecting our environment and wellbeing. Yet, at the operating level the programs and regulations have become vehicles for creating working tensions between management and bureaucracy. To gain a perspective on this uneasy alliance, it is best to look back in history.

In 1835, Alexis de Tocqueville, the French Aristocrat, toured the United States. The authors, Jeffery Beatty and Susan Samuelson, in their defining text entitled *Business Law and the Legal Environment*, quoted him as observing "scarcely any political question that arises in the United States that is not resolved, sooner or later gets resolved in the courts." We seem hooked on resolving large issues in the legal framework. Whimsically, our country was formed to protect us from our parent governments in England. Religious freedom and taxes (dumping tea in the Boston Harbor) are examples.

Our legal framework had its roots in the English Common Law system and even in our language. An example is the word sheriff. It is derived from the role of individuals called "shire reeves" (Beatty and Samuelson). Shires were the local English countryside communities and the shire reeves were the voice of the legal system. They collected taxes, mediated disputes, and even apprehended criminals. There was only a vague reporting to the central court system for their work.

When the French Normans invaded England in 1066, they instilled more discipline in real estate transactions 1) as a means of legitimizing land distribution by the Normans and 2) setting the precedent as a basis for common law. The earliest transactions were recorded as early as 1230.

Although the roots of the legal system were bound in history, the proliferation of laws and regulations as we know them today had a boost in the growth of our economy. The U.S. Constitution was written in 1787. It declared the right of the government to regulate commerce and provide for the safety and wellbeing to the country. A major example was seen in the legislation that arose from the early sweatshops that enabled the unions to seek organization and lobby for their membership. The legislation further grew when the Industrial Revolution occurred in the 19th Century. Similar boosts were seen in the Franklin Delano Roosevelt administrations of the 1940s as he drove for new agency development into economic crises the country faced following World War II.

Today we question whether the pendulum of government involvement has swung too far. Roger Trapp in an article published in *Forbes* entitled "Is It Time for Business and Government to Reconsider Relationship,"[1] argues that a new construct for how government and business act must be developed. Relying on government alone is not the answer. This theme is further developed in a book authored by William Eggers and Paul Macmillan entitled the *Solution Revolution*.[2] They cite examples like crowdfunding, ride-sharing, malaria in Africa, and traffic congestion in California. Trade solutions instead of just money grants may indeed fuel the solution revolution. Further, the report 15% annual growth of new ventures and even Fortune 500 companies in such a space. It is a hopeful scenario.

Specifically, we question whether the overall implication of the government intrusion in the business model has created a construct that compromises the probability of success of new projects and ventures. If enterprises around the world do not have these relationships, the balance is uneven and makes it harder to compete. A partial list of these relationships is shown in Figure 11-1.

[1]August 31, 2014.
[2]Harvard Business Press, September 2013.

Area	Regulations
Environment	EPA – Environmental Protection Agency
Workplace and Worker Safety	OSHA – Occupational Safety and Health Administration
Monetary	SEC – Security Exchange Commission
Tax and Financial	IRS – Internal Revenue Service GAAP – Generally Accepted Accounting Practices
Homeland Security	Multiple regulations
Industry Standards and Regulations	UL – Underwriter Labs CE – C Mark (European) ISO – International Standards Organization FDA – Food and Drug Administration FAA – Federal Aviation Administration
Governance	USPTO – United States Patent and Trademark Office SOX – Sarbanes-Oxley Laws
Other	Local codes for electrical, plumbing, and zoning

Figure 11-1. Government and business

It becomes tempting to look at the history of this involvement and trace its connections to modern enterprise. Certainly most of the rules were motivated by intentions of social good, environmental concerns, worker safety, and wellbeing. Even for those that are less altruistic in intent, one must ponder what the sum of the involvement does to the ability of a given project or firm to compete locally and globally.

I'm reminded of a story in my company that might serve as an example of this. Our projects were typically large in scope. The details were captured in project books consisting of three ring binders typically four inches wide. One project was in the United Kingdom. It was at a new Glaxo Pharmaceutical company lab in Stevenege, England. It took three years to secure the project. The project book for that effort was comprised of three such binders. No other project book in the company could match its details and complexity of regulations. The UK seemed determined to add layers of complexity.

It would be tempting to extrapolate this and suggest that the UK is handicapped in its ability to innovate and develop new products. Hardly the case, but one must wonder what impact the myriad of regulations has on the equations that control the probability of success of these ventures. The UK is years ahead of the United States and most other Western countries in their maturity of government interaction and regulation. The project books were just a visual representation. Is this detail any better and more relevant to this text? More important, how does it impact the probability of success of new and innovative projects?

Within the UK the environment is hardly static. Evidence of pendulum-like thinking was realized with the passage of legislation that enabled the creation of a Department for Business, Innovation, and Skills (BIS) in June, 2009. BIS allowed the merger of the Department for Innovation, Universities, and Skill (DIUS) and the Department for Business, Enterprise, and Regulatory Reform (BERR). The combined agencies embraced business regulation, company law, consumer affairs, employment relations, export licensing, higher education, and innovation. Changes are rationalized on increased efficiency and productivity. Political motivations are diminished. In the UK and other governments in the world, these changes are referred to as modifications in the Machinery of Government (MoGs).

As the forces of government interaction shift around the world, opportunities to innovate and successfully implement change move. China has long been a source of low-wage manufacturing. As their population and economy shifts to the cities, environmental and regulatory issues now appear as part of the landscape. Lower-cost manufacturing is shifting to other places in the Far East. The Chinese economy now boasts the manufacturing of airplanes and ships as part to its capacity.

To focus on any one global force may not be as important as recognizing the need for modern companies and projects to be "nimble" and "adaptive." These may be the next set of critical metrics of successful entities. What is intriguing is how early in the lifecycle these issues appear. Historically, they used to be issues of mature and later-stage entities. The timeline for adapting global alliances is long compared to the shortening lifecycles of new technological opportunities.

Global influences are tied directly to governmental interventions. There may be both positive and negative examples. There may be tax incentives, priority transfer options, and even direct subsidies for certain product and service areas. Equally, there may be restrictions and regulations that impact a given market sector. This is particularly true for technology-based products. An example of this comes from the automotive industries. For many areas certain safety specifications for items like the headlights, windshields, and bumpers were strictly applied to cars shipped to America. Most European and Far Eastern-produced cars could not meet these and relied on aftermarket organizations to provide them. The additional costs could result in the products being noncompetitive. There are stories of individuals visiting Stuttgart, Germany to pick up their Mercedes Benz cars only to be allowed to drive them in Europe (for tariff considerations) and then returning them to the plant for the cars to be remanufactured to the U.S. standards for an additional fee—amazing!

The Global Competition

Twenty-five years after the Russian Sputnik was launched, the United States convened a prestigious committee of academics and industry leaders under the stewardship of John Young, president of Hewlett-Packard. Its mission was to assess the state of America's ability to compete in the arena of global competition. The report was entitled "Global Competition, The New Reality." The date of publication was January 1985.

Briefly, the report concluded that America's "ability to compete in world markets must be improved." To "strengthen our competitive performance," we must:

- "Create, apply, and protect technology-innovation that spurs new industries and revives mature ones.
- Reduce the cost of capital to American industry and increase the supply of capital for investment.
- Develop a more skilled, flexible, and motivated work force.
- Make trade (exports) a national priority."

The report goes on to argue for incentives to create more public/private collaborations and to seek new models of commercialization. What is so amazing is that today's literature argues for these same directions. It then becomes important to recognize what has changed since that report. The rate of change of communication (via the Internet) and the advances in enabling technologies that have been noted in this book have increased at breathtaking speeds. The global population has increased from 4.5 to 7 billion (see Figure 11-2), natural resources have dwindled, and the balance of trade has shifted from Europe and the United States to the Far East. All of these factors impact the probability of success in the projects and the new enterprises we might contemplate.

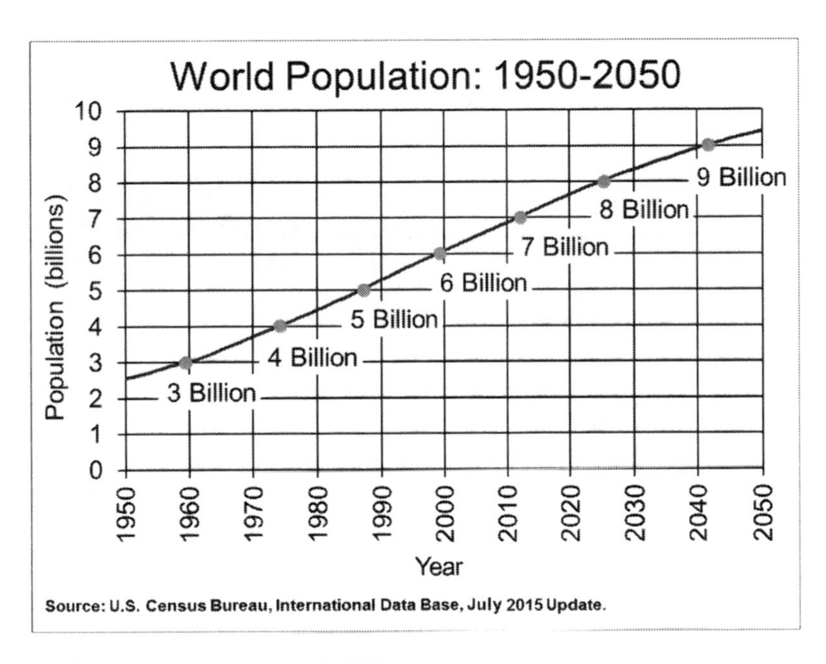

Figure 11-2. World population 1950–2050

Perhaps the most troubling part of this dialogue is that the ability of the United States to fund new innovations and embark on new trade initiatives has diminished remarkably. At the time of the commission's report (1985), the gross public debt was a nominal $2.5 billon; today it approaches $15 billion. Servicing the debt alone becomes monumental and compromises the ability to fund new directions (see Figure 11-3).

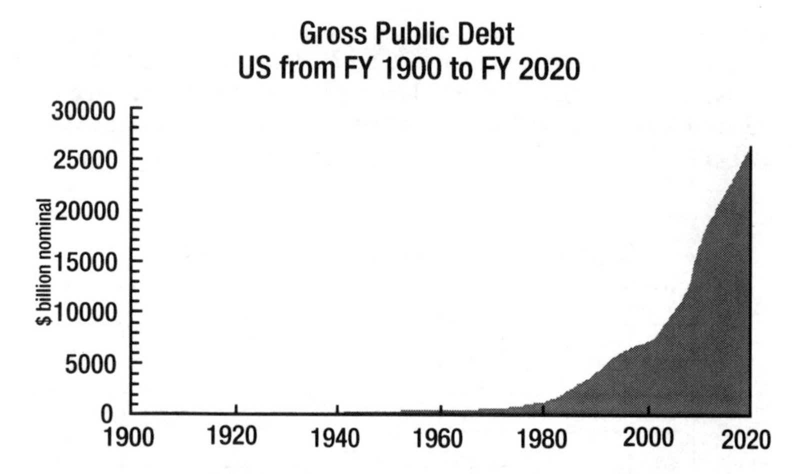

Figure 11-3. U.S gross public debt[3]

Each year, the Battelle organization and *R&D Magazine* publish a Global R&D Funding Forecast. In the 2014 edition, they reported that U.S. R&D funding rose by 1%. In the same time period, China's increased spending by 6.3% and Europe by 4.6%. In combination, these three areas spent 87% of the world's R&D for a level of $1.6 trillion. If the year-to-year trends continue, the Asian countries, including China, will outspend the United States by 2018. Within the U.S. category, 70% of the R&D spending comes from industry.

In the narrative of the report, it states that "R&D is a long-term investment in the future that serves as the cornerstone for innovation-driven growth." It then elaborates about the importance of an ecosystem to utilize the benefits of R&D. The ecosystems allows the benefit to "stick" until commercially viable. Elements of the ecosystems include:

- Large investments in human capital to ensure a talent pipeline of the required skills. The importance of STEM programs was emphasized.
- Science is partnered with commercial visions and entrepreneurial efforts to allow advancements of the efforts.
- Capital is available for all stages of effort from R&D through proof-of-concept to the final product.
- Government support is established and responsive to industry collaborations with academia.

[3]Source: `www.usgovernmentspending.com`.

When the report investigators asked leading researchers about the concern in affecting future trends, there was a surprising finding in that global trends such as natural disasters and renewable energy sources were high on the list.

To extend the issue further, measures of productivity reveal the potential for global growth. In a report from the McKinsey Global Institute entitled "Global Growth: Can Productivity Save the Day in an Aging World?",[4] it was indicted that we have enjoyed a period of 50 years of growth as measured in GDP. The rate of growth changed from 2.8% per capita growth to 2.4%. Although this change seems small, it is further complicated by the number of emerging countries that help to drag down the overall performance. Nigeria, for example, would have to boost their productivity by 80% to catch up to the global numbers. The report focuses on agriculture, food processing, automotive, retail, and healthcare. It was relatively optimistic that we have not run out technology and innovation that could be utilized to improve our global productivity but question whether we have the governmental and industrial will to adapt to the best practices we will need.

At a macro level, trends appear that are buffeted by local economies, currency, and political changes and the dynamics of emerging economies. Through the openness of the Internet and global transport capacity, the forces impinge on new opportunities and can diminish the probability of their success. On the other side, global shifts can also present opportunities in terms of collaboration and business exchange.

An example of this was conveyed to me by a former executive of the Dennison Corporation of Framingham, MA. The company was involved in the manufacture of pricing tags for the retail industry. In the clothing goods sector, they observed that particularly high-end products were often designed in Paris or Milan, manufactured in the Far East, and finally sold in American distribution channels. Cataloging and displaying consumer tag information accurately and in a timely manner became daunting because of the many hands and disparate locations that the products and services utilized in the goods manufacture of the products. They developed a process where the stages of tag information could be generated in real time in the disparate locations. The overall productivity (and error losses) were quite positively impacted. Of course, there are many other examples of this type of global interaction. What is so intriguing is that they would have been unavailable just a few years ago, before the technologies used to accomplish them were available.

[4]McKinsey and Company, 2015.

The Local Context

In Figure 11-1, we started to look at how regulatory constraints and compliance requirement can impact even the earliest of commercial projects. It is appropriate to look at some of these details more closely. A starting point might be just to look at the list of agencies and rules that impact any organization, including the startups. The agencies are broadly categorized as Federal and Local (state and municipality).

Perhaps an entry point into the world of governmental involvement might be the Small Business Administration (SBA). Started in 1953 by President Dwight D. Eisenhower, it had two sweeping mandates. The first was to administer multiple levels of government-backed loans to small business for both the capital and growth projects. The second direction was to "aid, assist, counsel, and protect, insofar as possible, the interests of small business concerns."[5] It went through many iterations and was buffeted by both Republican and Democratic forces. The Democrats wanted to enlarge its scope, while the various Republican administrations went as far as trying to abolish it. Today, under the Obama administration, there is movement to bring the SBA to presidential cabinet visibility. In addition, in December 2010, President Obama created the Small Business Jobs Act to not only allow an additional $30 billon in lending programs but also to provide up to $12 billion in new tax cuts for smaller enterprises.

SBA is huge. It is comprised of 22 separate offices with interest ranging from entrepreneurial education to international trade to veterans' business and women's business ownership. Within its broad reach, it also helps administer SCORE, which is the Service Corp of Retired Executives. SCORE is a mentor group comprised of 350 chapters around the country. The SBA has at least one office in each state, approximately 900 Small Business Development Centers located in colleges and universities, and 110 Women's Business Centers to assist in its outreach into the early stage community. Some of the specific outreach initiatives it sponsors include (not in order):

- Intellectual property. The issues surrounding patents, copyright, and the myriad of issues surrounding "the freedom to operate" in a litigious society.
- Environmental regulations. Deals with multiple regulations involved with not only working with material for manufacture but rules dealing with the full cycle of all materials.

[5]https://www.sba.gov

- Foreign workers. The laws dealing with employment eligibility and the ever-changing immigration rule as the political winds surrounding them change. Included are the issues surrounding the search for the best international talent, which is rigidly controlled by Visa regulations.
- Employment and labor. The specific rules that surround the hiring of general workers.
- Business law. From the basic of the Uniform Commercial Code (UCC) that regulates contracts and business transactions to specifics such as those that regulate commerce on the Internet.
- Financial laws. Accounting transactions are subject to multiple laws, starting with the Internal Revenue Service (IRS) to the Securities Exchange Commission (SEC) for public equity transactions to full accounting transaction practices under the Generally Accepted Accounting Practices (GAAP).
- Regulations and permits. There are multiple codes and permits for operations that appear on the local level that embrace plumbing, electricity, and other municipal services.

Even if these interfaces are proper and help regulate the flow of commerce, each interaction, whether small or large, costs time and human resources. If projects are to realize their potential, the additional burdens must be borne by the enterprise. At minimum, a certain level of marginal or more fragile projects are declined. What would have happened if certain ideas, ranging from the cotton gin to the automobile to the television and the phone, had been subjected to the levels of scrutiny now employed? Some of these technological innovations helped define America in global markets.

If this is extended to a global perspective, the rules are not designed for parity. Certainly the environmental concerns and regulations in China are trivial compared to the United States and Europe and are grossly imbalanced. To offset this, there are multiple governmental and political initiatives that seek parity among the countries. Global carbon tax efforts that normalize global shifts in the environment are an example. Agreements such as these are certainly subject to political pressures. One example is the Kyoto protocols, which were essentially boycotted by the United States. Without America's participation, the impact of these environment laws was nullified. Until the agreements are secured, each element of government intervention establishes commercial imbalance. These imbalances are registered directly on the potential profitability and capital requirements of doing business in each sector. One has to wonder how much consideration is given to the various trade imbalances that are created when these laws and regulations are crafted.

There are offsets and new opportunities created in this implosion of regulations and laws. One might be seen in the case where the Environmental Protection Agency (EPA) promulgates new emission standards for heating plants. Older technologies such as coal-fired open stack heaters are being replaced with more efficient and cleaner technologies such as natural gas and exhaust stack scrubbers. Clearly these are better for the environment and maybe even more efficient in the delivery of heat to a given facility, and the conversion deflects the use of capital that might have been used to create new enterprises. That new commercial opportunities for natural gas heaters and scrubbers might be rationalized as the benefit. Even if this is so, the overall productivity is compromised.

Drilling Down

Governmental and regulatory influences on early stage projects seems to fall into three categories. They include environmental concerns, worker rights and safety, and regulation of commerce. Each category has its roots in promoting social or environmental concerns and certainly can be justified in this basis. Specific areas of the world such as Europe and Scandinavia are known for the presence of strong and invasive interactions. The Far East and China in particular are at the other side of the balance argument and currently have less stringent rules. Each of these areas is undergoing constant change. It is a moving target. There is a long list of global trade and environmental commissions and treaties seeking to normalize them. Clearly it is a fluid set of balances and interactions.

If the new world of opportunities ahead embraces a global competitions, the range of interactions is large and at any one time seems to be imbalanced. Given that, early stage projects seem increasingly vulnerable to a set of constraints that are well beyond their capacity to meet. Let's look at the specifics of a startup manufacturing entity in the three dimensions of environmental regulation, worker rights, and safety and commerce regulation.

In each category, there is at least one dominant regulatory agency. In the American environmental space, for example, the U.S. Environmental Protection Agency (EPA) is the dominant force. Created by executive order under President Richard Nixon in 1970 and later ratified by the House and Senate, it has broad regulatory and enforcement powers in the space of air and water quality. Its scope embraces the air, water, land, and endangered species and hazardous waste regimes. It has spawned endless regulations and governing laws that impact both small and larger entities.

In a report entitled "The Impact of Regulatory Costs on Small Firms" by Nicole and W. Mark Crain[6] a significantly larger share of compliance of the environmental regulations was borne by smaller companies. This is shown in Table 11-1, published in the report.

Table 11-1. Distribution of Environmental Compliance Costs by Firm Size in 2008

	Cost per Employee			
Type of Regulation	**All Firms**	**Firms with <20 Employees**	**Firms with 20-499 Employees**	**Firms with 500+ Employees**
All Federal Regulations	$8,086	$10,585	$7,454	$7,755
Environmental	$1,523	$4,101	$1,294	$ 883

What seems so astonishing is how skewed these costs are with respect to the size of the companies. Those earlier stage organizations that we rely on to be innovative and become the basis for successful commercialization of technology are hit hardest. As Table 11-1 further suggests, this disproportionate burden is also borne by smaller companies in all governmental regulatory categories.

To look at anther slice of this disproportionate balance, we can look at an accumulation of economic forces such tariffs, fees, and taxes to see similar patterns in the category of all Federal regulations. Although beyond the scope of this report, there is another layer of local and state costs such as taxes and permit fees that must be added to the tally. Many of the costs are "fixed" expenses that must be borne independent of the number of employees. These costs tend to hide in the accounting models and are not listed as line items.

There is clearly a balance between public and private allocation of funds. If a corporation invests in a given environmental technology to improve its productivity technology, that is one category that may give it commercial advantage. If the investment is made to satisfy an EPA regulation, that is another. That decision is more that an accounting convenience, but rather is a major force that affects the company's competitive position.

Not only does this burden fall on companies least capable of accepting the financial and cash flow burden, but it sets up an imbalance in the competitive global and foreign markets. Technology and innovative solutions alone cannot provide the balance needed for any one country to compete. The basis for many regulatory elements may indeed have altruistic and sustainable arguments. They certainly a based in political forces. Somehow there must be

[6]Small Business Administration Office of Advocacy, September, 2010.

equilibrium of forces to compensate for these imbalances. If China or a third-world country can dump products into the U.S. economy that have minimal environmental effects, it does seem to be unfair economic element.

Putting It Together

In this chapter, we looked at the many ways the various federal, state, and local governmental organizations impinge on the modern corporation. The rate of involvement is unprecedented. Although the basis of many regulations seems altruistic, their purpose and scope are disproportionate in the way they impact smaller entities. This puts a burden on the early fundraising efforts and the resultant dilution of equity, and it also limits the use of capital that can be applied to the organization's growth purposes. What is more compelling is how powerful the environmental and economic constraints are around the world. In addition, they also change. Combined with the disproportionate regulatory burden on the early stage company, one wonders what the ability of any one country to innovate and recognize the benefits of commercial activity will be going forward.

To be sure, not all regulations bring with them financial and human resource obligations. Some are actually additive in nature. Examples include reduction of certain tariffs or Small Business Innovation Research (SBIR) grants that allow collaboration between academia and industry partners. There might be local and federal incentives to provide solar energy. There might be available job training or job creation incentives. These programs prevail in all aspects of an ongoing enterprise. Cleary those incentives must be incorporated into operating plans and utilized when possible. If they provide a global competitive edge, then they must be employed. Whether the incentives are driven by social, political, or altruistic motivations, they are in a state of constant change. The management of the new technological-based enterprise requires a level of nimbleness and adaptation that has never been required before to take advantage of them.

Looking Ahead

With the rapidly changing landscape of elements in the commercialization of technology, it becomes uncertain to forecast what parts will have the most significant impact. In the next chapter, however, we will look at new trends of investments and opportunities. In addition, we will look at the changing roles of incubation, patent law, market communication, and the Internet. We will also look at how teams are formed. In addition, we will examine these changes in a global context. For sure, it will be an exciting time ahead.

CHAPTER

12

Looking Forward

Throughout this text there has been a repeated focus on the rate of change of the elements in the commercialization cycle. These changes altered the enabling parts and helped us do things more effectively and faster. New means of funding early stage projects, new forms of incubation, more effective teaching, and better modeling tools are just some of the items to be noted. With these dramatic changes there is an increased risk of forecasting. But to codify them as they exist today becomes an interesting task.

A Changing World

To look forward certainly carries its own caveats about uncertainty. One aspect that we can easily agree on is that there is a tremendous rate of change surrounding the commercialization, entrepreneurship, and innovation cycles. We can at least see directional change by looking at the several categories that surround change. They include:

- Early stage investments
- Incubation trends
- Global competition
- Technology and IP changes
- Entrepreneurship training
- The Internet and customer demand
- Economic demands

Early Stage Investments

In 1946 Georges Doriot (former Dean of the Harvard Business School and sometimes referred to as the "father of venture capitalism") started the first recognized venture capital firm called American Research and Development Corporation (ARD). Its landmark early investment was $70,000 in Digital Equipment Corporation. With that investment a new investment industry was born that is estimated at $25 billion each year. The relevance of investments like these is not that they enable hundreds of early stage ventures to start but also that they allow underperforming portfolios such as pension funds to place a limited amount of their assets in high-risk/high-reward, professionally managed projects. They are the folks who become the limited partners of the venture firms. The early venture capital model carried the assumption that there would be a robust market for exiting their investments in initial public offerings (IPO) and the allure of high stock price multiples.

That speculative model carried industry growth until the year 2000, when the infamous "dot com" bubble of Internet-based stock prices collapsed. An abrupt collapse of perceived value created significant loses, including those involved in venture capital investments. The industry response was to retreat to more established (less risky) investments. This left a significant gap in early stage investments and in startup companies.

In 1978 Professor Bill Wetzel (faculty member of the University of New Hampshire and founder of its Center for Venture Research) observed the behavior of individuals placing private investments in early stage ventures. Most of them were located in New Hampshire. These individuals had cashed out from successful technology-based companies and invested in early stage projects. He coined them "angels" from the model of Broadway show investors who take investment risks before shows are launched.

Today the annual level of angel investing equals that of venture capital efforts. Angels place less money per deal but participate in more deals. They also participate in earlier stage deals and those in the "gap" left behind in the movement of venture investors to later stage deals. Their comparative relationships are shown in Figure 12-1.

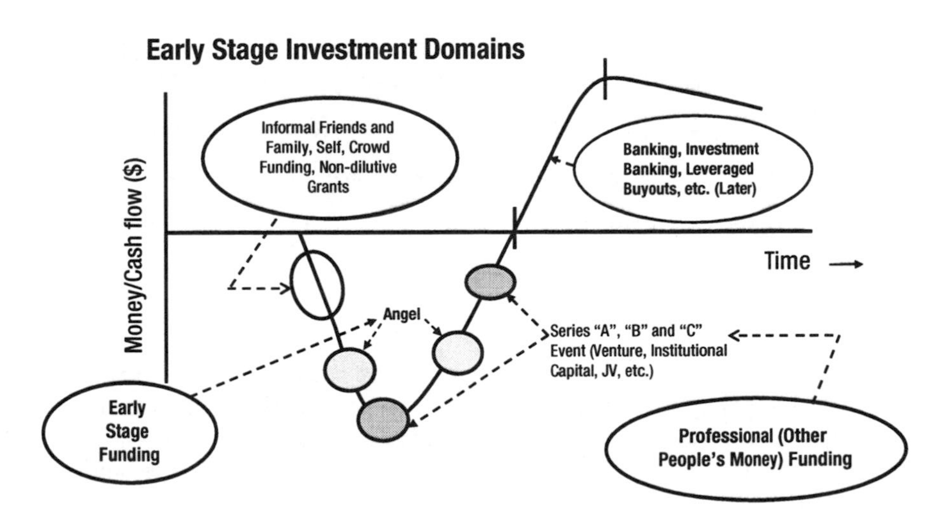

Figure 12-1. Early stage investment cycle

Both venture and angel deals have a common issue of who is capable of making those investments. The Securities Exchange Commission (SEC), in an attempt to protect those unable to absorb risk because of limited means, defined the "accredited investor." Both angel and venture capital investors must meet the test of the accredited investor—they must have a Net worth of $1 million and an annual income of $200,000. Primary residences are not included in this tabulation. Later, the Dodd-Frank Act extended this definition. This means that only individuals with significant worth can invest.

On April 5, 2012 President Obama signed a bill entitled "Jumpstart Our Business Startups Act" or the JOBS Act for short. The act enables new sources of capital to be made available to early stage companies. It does so by allowing equity issues to be traded on the Internet and reduces the qualifying level of net worth of individuals. It also allows for an increased number of shareholders that could be brought into an organization before they had to be reported to the SEC. The JOBS Act democratized early stage investment, provided for a new distribution of wealth, and opened social media pathways for a new type of funding called "crowdfunding." Almost anyone could invest in a company from anywhere in the world.

It also spawned a new crowdfunding industry characterized by the emergence of a company called Kickstarter. Started on April 28, 2009 by Parry Chen and others, Kickstarter became a portal where outsiders could invest in their portfolio companies. According to the web site in 2012, Kickstarter has raised $1.5 billion for 207,135 projects and touts a 40% success rate. Its model is further defined as an assurance contract whereby it will not release the funds for a given project unless all of the money has been raised. Kickstarter requires a 5% fee for its effort. It has certain restrictions to its projects that include:

- Banning the use of photorealistic renderings or simulations
- Banning genetically modified organisms
- Limiting items to "sensible sets"
- Requiring a physical prototype
- Requiring a manufacturing plan

Kickstarter is not alone. There is an array of companies like GoFundMe, Indiegogo, Teespring, Patreon, Crowdrise, and others that are already sharing the space. It is still unknown how effective they will be and how this new form of capital acquisition will survive, but what is clear is that the landscape is rapidly changing.

One more dimension of this new trend is the addition of non-equity based crowdfunding. In this scenario, goods (or a future promise for them) are exchanged for cash. An interesting example appeared in the April 29, 2012 issue of *The New York Times*. The article describes the journey of a 26-year-old engineer named Eric Migicovsky. Eric had developed a line of wristwatches called Pebble Watches that conveyed text messages from iPhones. First he tried the traditional venture capital routes to raise startup capital but was flatly turned down. He then turned to Kickstarter. He offered a Pebble Watch for anyone who would contribute $99. No equity was exchanged. In less than a week, he raised $7 million from nearly 50,000 people. Later company reports indicated that a total of $10 million was raised by the company in less than the two-week window offered by Kickstarter.

Pundits of this process say that without appropriate vetting, committee review, proper due diligence, and strict term sheets, the upside investment quality would falter. The same was stated about angel capital when it first started almost 40 years ago. Today, the angel investing industry has its own national best practices organization (ACA) and is reported to have placed an estimated $24 billion in 2014. This amount is equal to the funds placed by the venture capital industry in the same year.

Incubation

There is a moment in the early lifecycle of an organization in which the project is quite fragile. Teams aren't complete, technology has not matured, or funding itself might not be totally secured. All of this is independent of the perceived value of the enterprise. Successful companies certainly prevail. Some need more time and the resources to achieve success.

In 1959, the first recognized incubator to meet these needs was created in Batavia, New York. The idea of a public entity to achieve this help took years to develop. In 1980, there were only 12 recognized "business incubators" in the country. As more industrial plants closed in the Northeast and the economy seemed to change in the 1980s, more incubators grew under the rubric of economic development. In 1982, the Ben Franklin Partnership Program was created as a first statewide organization. It became a model for many others. In 1985, an organization called the National Business Incubation Association was formed to serve as a clearing house of information and best practices. It was comprised of 40 members. Today there are 1,600 members.

In more recent history, the energy to create incubators reached beyond economic development terms that promised the creation of jobs through the creation of new companies. Internet companies that no longer require large capital investments in laboratories and physical resources have emerged at a breathtaking rate. A simple working space, administrative help and a broadband Internet capacity seem all that was needed. The Federal government did not stay on the sidelines. The U.S. Small Business Administration (SBA) created Small Business Development Centers (SBDC) to assist early stage companies. There are about 900 centers in the country. Other government programs such as the National Science Foundation (NSF), I-Corp, and the Small Business Innovation Research (SBIR) are further examples.

The incubation process is indeed global. In Switzerland, for example, there are government/industry cooperatives like the Technopark in Zurich and the Blue Factory in Fribourg. In Israel a state-run system of incubators was launched by the country's Office of Chief Scientist in 1991. Its goal was to enable a "startup nation" in the country. In their model, the government contributes 85% of the early startup costs. It is not without its challenges. In the May 19, 2015 issue of the Israeli daily newspaper called *Haartez*, there was a lead article that asked, "Technology Incubators: Has Their Time Passed?" Although, the director of the organization, Yossi Smoler, commented that the model is changing to one of more inclusion of corporate and financial support, the article pointed out that open source software and cloud computing has changed the way many new companies are formed. Rather than join in this dialogue, I wonder if newer models need to be created.

There are already new models in operation. In Boston, for example, an accelerator program called Mass Challenge was created about five years ago. With international representation around the world, it claims to be the "world's largest startup accelerator." It offers a large itinerary of coaching, mentorship combinations, and access to investors and other resources.

In the February/March 2013 issue of the *NBIA Review,* published by the National Business Incubator Association (NBIA), the new director of the organization, Jasper Welch, is interviewed about his thoughts of the future of incubation. What is clear, he states, is that incubation must reach beyond the bricks and mortar model to operate in the "next practices." Others have referred to this space as the virtual incubator. What is again clear is that this area of incubation will have to change with the changing needs of early stage companies.

Global Competition

If there is one theme that captures the essence of global competition, it is that of change. The United States and Europe have long dominated the expenditures of R&D investments. Yet, in a 2014 Global R&D funding forecast published by *Battelle and R&D Magazine* (rdmag.com), December, 2013, they stated that Asian countries have grown their share of global spending on R&D from 33% to 40% in the past five years and that "Southeast Asia has become the world's largest region for new research investments-a trend expected to continue through the decade." The shift in knowledge production is significant. China's investment in education has borne fruit. The "country has overtaken the United States in the number of doctorates awarded in science and engineering. They have 1.6 million researches and academic and 30 million students enrolled in higher education institutions." Once a bastion of low labor costs, China has been motivated by these trends to outsource labor costs to Africa, South America, and the Middle East. One byproduct of their educational and research activity is that since 2011, China has accounted for the greatest number of patent applications globally.

This shift in technology development has an interesting byproduct. In China, India, and Brazil, the number of the world largest corporations in residence has grown 6.7%. In the same time period, the United States, United Kingdom, and Germany have reported significant declines.[1] One further indicator is the number of Total Early Stage Entrepreneurial Activity (TEA) index. TEA measures the percentage of individuals in an economy who are in the process of starting or running new ventures.

In the five countries reported, the United States and European Union reported the lowest TEA numbers, while Latin America and Africa reported the highest (*Global Entrepreneurship Monitor* (GEM) 2013 Annual Report, Jose Enesto Amardo, Amanda and Niels Bosma; Babson College et al., 2014). Innovative entrepreneurship is defined as creating a product or service that represents significant commercial opportunities. The presence of supportive environments in the emerging countries includes increased access to funding for early stage entities (venture capital), changes in culture, supportive regulatory initiative (such as eliminating capital gains taxes), relaxation of rules restricting foreign investments, and educational systems integrated with these trends.

---

[1]The Super-Cycle Lives: Emerging Markets Growth Is Key," November 6, 2013, `https://www.sc.com/en/news-and-media/news/global/2013-11-06-super-cycle-EM-growth-is-key.html`.

Technology and the Changing Landscape of Intellectual Property

Intellectual property protection manifests itself in the form of patents, copyrights, trade secrets, and licensing innovative ideas. In common they provide a barrier to infringement and thus make it more attractive for financial investment and talent acquisition in those same ideas. The common law beginnings of intellectual property protection envisioned just this. What they didn't see was the explosion of technology, the Internet, and the global competition that would surround it.

We have moved a long way from the lone inventor hunched over his or her worktable and shouting "eureka" as the proverbial light bulb of invention went on. Today we are involved in more collaborative complex technologies that vary across business sectors. Certainly, genomic models and their discovery, which are at the basis of modern medical science and the rapid advance of software and Internet-based solutions, were not envisioned by those who crafted patent laws around the world.

In 1970 a group of United Nation member nations created the World Intellectual Property Organization (WIPO) to "promote and harmonize intellectual property law internationally." Among its early initiatives was the promotion of IP from industrial to developing countries. Creating a global context for IP is a daunting role. Cultures, trade agreements, and local judicial systems all complicate the drive to broaden encompassing laws. Yet, the needs for these agreements is greater than ever due to the global nature of technological competition. A newer initiative entitled the "Trade Related Aspects of Intellectual Property Rights" (TRIPS) within the WIPO is attempting to find flexible, yet encompassing directions of new legal development. A "one size fits all" direction seems an elusive goal.

At least one other consideration that affects the "flexible" approach that the TRIPS effort is trying to accomplish is that of shorter terms for the IP protection. In earlier years a 20-year term of patents seemed reasonable as a protective envelope to commercially recognize its value. Rapid technological advancements, improved enabling tools, and shorter lifecycles have challenged the 20-year boundary to the point where the term actually is perceived as a barrier to innovation and change. The value proposition simply becomes outdated before the patent or copyright expires.

Sometimes IP thwarts both social and societal need. An example is in the field of pharmaceuticals. Patents give their owners the right not to commercialize a given technology. It would be hopeful if societal, social, or even sustainability issues could be brought to bear as mitigating issues. At this point there seems little effort being exerted to find such solutions. In a telling article written by Richard Spinello, who is on the faculty of Boston College (entitled "The Future of Intellectual Property"—Ethics and Information Technology, Euwer Academic

Publishers, Netherlands, 2003), he argues that the same issue applies to Internet-based IPs. In the absence of succinic rules of protection and privacy around the IP, the very basis of the innovation that created the application is in jeopardy. The World Economic Forum's 2014 meeting in Davos wrote a member's position paper entitled "Rethinking Intellectual Property in the Digital Age." It proposed a series of "guidelines" that encourage collaboration—open dialogue, consumer engagement, and development of "shared goals." Although not as specific as the incentive to modify laws and regulatory mandates, it may set the stage of additional cooperation.

One last but disquieting theme impacting the future of IP is the emergence of patent "trolls." They are firms that acquire rights to patents but do not manufacture any products or services. An example is a troll firm buying up the patent rights of a bankrupt company for the purposes of suing a more innovative and successful company and gaining instant fines derived from the suit. In the United States, there is the concept of share legal expenses, so the risk of the challenge is minimal (or shared). In the November, 2014 issue of the *Harvard Business Review,* James Bessen, an economist at the Boston University School of Law, notes what he deems is an alarming rise in the number of patent litigation cases created by these trolls. This trend is illustrated in Figure 12-2.

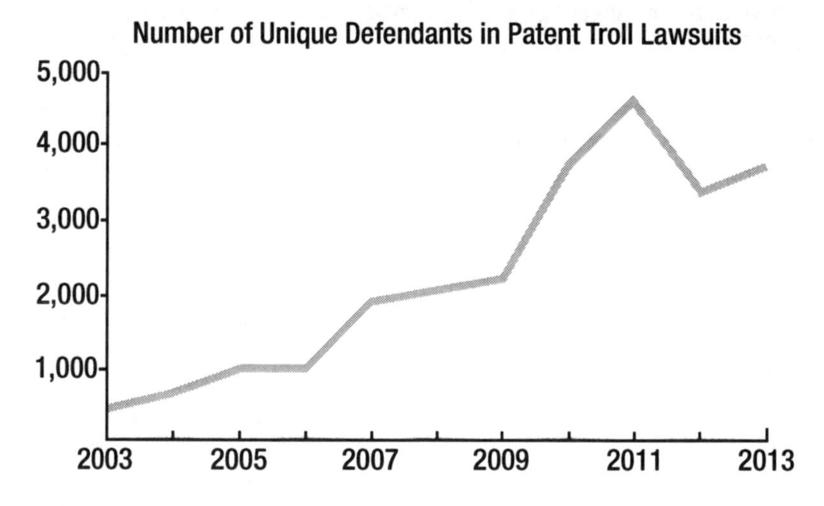

Figure 12-2. The rise of patent litigation cases created by Internet trolls[2]

Further in the same article Bessen refers to a study conducted by Roger Smeets of Rutgers University and Catherine Tucker of MIT, where Smeets observes that smaller companies' R&D spending diminished significantly when

[2]Source: HBR.org.

sued by the trolls. Tucker reports a 14% decline in venture capital firms in those same companies, while Smeets cites a 19% decrease in R&D spending by those same companies. Both trends are shown in Figure 12-3.

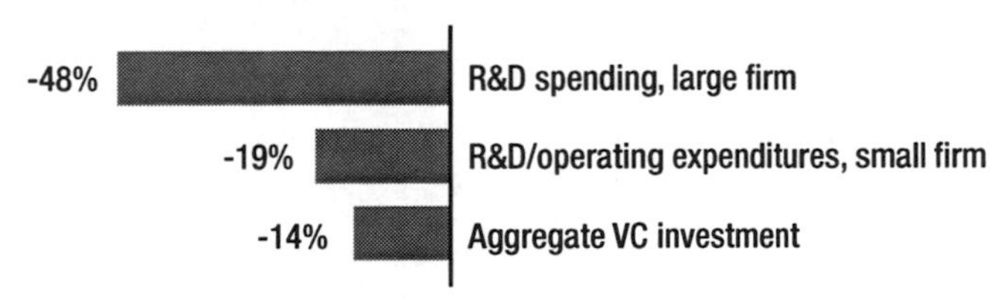

Figure 12-3. Effect of patent troll lawsuits on innovation[3]

Whatever the short-term impact of troll's litigation prowess is on the innovative expenditure, it is only the federal government's ability to legislate those changes in the laws that will mitigate the negative effect of their disruptive activity. Those changes and the ability of global entities to collaborate on their regulatory efforts will become markers for future changes in IP. Clearly the new forces of technological global innovation and the emergence of the Internet on music, film, and social media will be driving forces for these changes.

Entrepreneurship Training

One of the major historical bastions of the American economy has been the emergence of the entrepreneur. In many cases, he or she altered the landscape of our sense of business. Examples are many.

FedEx was started as a term paper in 1965 at Yale by Fredrick Smith. The paper was perceived by Smith's professor as non-revolutionary. Today, FedEx is a firmly established fact of life. Michael Dell started the now famous Dell Computer company in his dormitory room at the University of Texas in 1984. Bill Gates was at Harvard when he and Paul Allen conceived the Microsoft operating system as a business.

In an essay entitled "The Chronology and Intellectual Trajectory of American Entrepreneurship Education,"[4] Jerry Katz, a professor at Saint Louis University, cited the first economics/business courses references as early as 1876. The first published business text was cited at Gordon Baty's "Playing to Win."[5] From there, an accelerating pace of symposia, journals, and case studies appeared. Business schools appeared but cantered their offerings in management. Entrepreneurship seemed to be a minor offering.

---

[3]Source: HBR.org; Research by Catherine Tucker, Roger Smeets, Lauren Cohen, Umit Rurun, and Scott Kominers; Analysis by James Bessen

[4]*Journal of Business Venturing* 18, no. 2. (2003): 283–300.

[5]Reston Publishing, 1974.

Looking ahead, the Kauffman Foundation published an article in February, 2015 written by Jason Weins and Emily Fetsch entitled "Demographic Trends Will Shape the Future of Entrepreneurship." It cited the growth of millennials and baby boomers as the primary driving force. Education has responded by offering in 250 courses in entrepreneurship in 1985 to 5,000 in 2008. These demographics are in Figure 12-4.

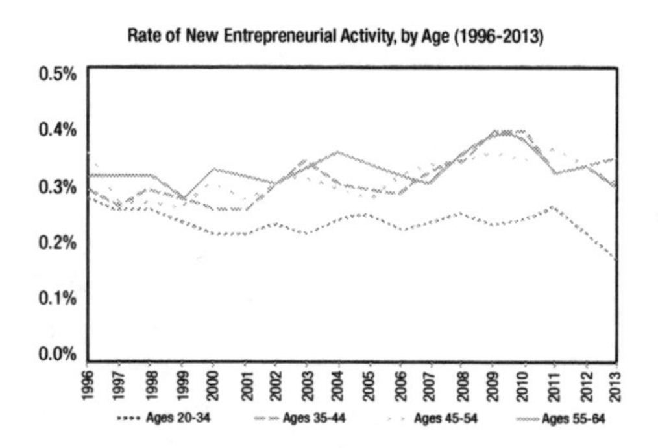

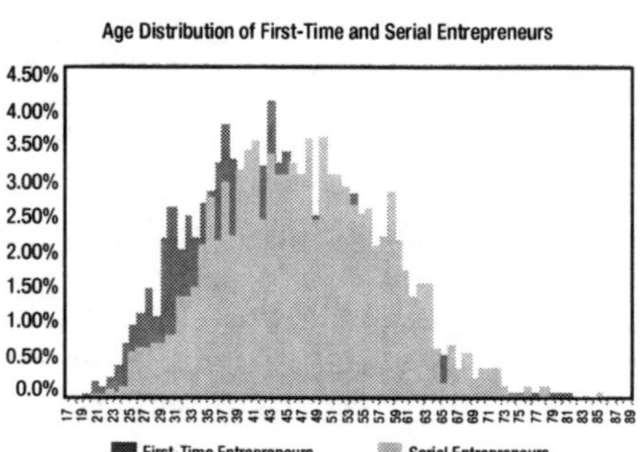

Figure 12-4. Entrepreneurship demographics[6]

This means that new high-tech startups will be twice as likely to be started by individuals over 50 than those younger than 25. That shift may not be enough for the United States to maintain its leadership in entrepreneurship. Jim Clifton, Chairman and CEO of Gallup, wrote in the January 13, 2015 issue of the *Gallup Business Journal* that for the first time, the United States has projected that the number of business closures will exceed the number of startups. The United States will be twelfth in the global lineup of rates of startups. To reverse this trend, he argues that to "get back on track we have to quit pinning everything on innovation; we need to start focusing on the almighty entrepreneurs and business builders."

Academia is responding to these changes in multiple and notable ways. It was only 40 years ago that the first master's in business administration (MBA) with an entrepreneurship concentration was offered by the University of Southern California. Today there are many programs in place in cache universities such as Harvard, MIT, Stanford, Babson, and Kellogg. They are making significant inroads to teaching students and developing sufficient staff to teach them. Novel experiments in case writing have spawned new teaching tools of simulations and online courses. For a variety of reasons including outreach, student enrollment, and spiraling costs, new means of delivering information are now

[6]Sources: Robert W. Fairlie, Kaufman Index of Entrepreneurial Activity (2014), using data from Current Population Survey, US Census Bureau; Kauffman Firm survey.

in place. Blended learning, for example, is where students can continue their education remotely but still maintain presence on campus several times a semester. Online courses are proliferating and collaboration with industry, including courses taught in remote plants, are currently in place.

Supporting these trends are an endless list of business plan competitions, entrepreneurship clubs, incubators, and practice labs on campuses. Courses integrated as minor offerings in alternative fields such as engineering and science are also prevalent.

The Internet and Customer Demand

To forecast the future of the Internet is a bit like asking the Wright Brothers what they thought of intercontinental flying at altitudes 40,000 feet and almost 600 mph. The rate of change and the acceptance as working tools in applications as diverse as medical information, marketing, finance, and entertainment are simply awesome.

The Internet as we know it today was not a single invention. Its first enablement was the concept of packet switching developed for the Department of Defense. When groups of data could be sent, the application of Internet Protocol allowed those packets to be sent between users. Tim Berners-Lee added the concept of a the World Wide Web to the process, which allowed e-mail, instant messaging, and finally voice over capability between users. All this occurred in just 30 years. Today we just accept its capacity in a verity of devices like handheld smart phones, financial interaction, and text and photo sharing, to name a few. Best of all this capacity is being recognized throughout the planet.

There are some early clues as to what the future holds. Let's look at three. First, offerings of the "cloud" promise better access to data through multiple devices such as pads, computers, smart phones, and watches. It allows better allocation of resources because individual users do not have to invest in mainframes and memory that becomes easily obsolete. It is global so worldwide timing issues and Interfaces become invisible to the users. The cost of the services can be allocated on a pay-as-you-go basis. This aspect alone is having enormous impact on the costs absorbed by the users.

There is a least one more consideration that affects the use of the Internet going forward. It is the speed that the technology can transport information. Today, the average home broadband speed is 10 megabytes per second (Mbps). Throughout the United States, this number varies wildly. In the populated centers like New York and Washington, speeds start to approach 100 Mbps. It is projected that by 2020, the number could reach 200 Mbps. Today, the infrastructure of fiber optic cable and transmission equipment is expensive and improvements are hard to justify financially. One possible solution is to rely on government legislation and funding to normalize this. Whether that happens is interesting to conjecture. Currently the United States is far behind countries like South Korea and Sweden in their ability to provide fast service to broad markets and applications. With increased capacity, more elaborate and complex information can join in the use of the Internet.

So here we see a technology that promises significant increases in its ability to transmit information, an exploding world of applications, apps, and devices, wireless connections, and the almighty cloud as alternatives for bringing all that information to us—wow.

All of this is not without its issues. Clearly at the top of the list are the domains of privacy and security. Hardly a day goes by without hearing that some system has been hacked or has been infected with malware, a Trojan, or some other virus. Some of this is a footrace between protection software and the worlds of individuals determined to thwart the use of the Internet. When domains of medical records, financial data, security, and other sensitive materials are considered, this issue becomes more important than ever. Firewalls and encryption schemes to protect the data are certainly defenses against security intrusion, but absolute integrity seems an elusive goal. I don't doubt that as a society we will also adjust to a greater degree of intrusion. Just look at the current offerings of Google and Facebook to see how much "private" information is now in the public domain. Tracking software "cookies" can follow most transactions we do on the Internet. The fields of cybersecurity will grow as we find new forms of data protection.

No matter which way we view the future of the Internet, its overall potential is simply awesome and will impact how we commercialize new technology and its applications.

A View from the Trenches

A book like this one would be incomplete if it did not offer a perspective from the entrepreneur's view. Every element of an entrepreneur's environment is subject to change. It is not cool to remain a startup! To cite a few change elements:

- The renewed emphasis on innovation and entrepreneurship at all levels of the government and social frameworks. Endless federal and local programs and funding are available at a rate never imagined.
- Educational systems including new courses, seminars, workshops, and student-run clubs and laboratories are increasingly prevalent. Most academic environments encourage this activity.

- Available sources of capital in the form of new structures like crowdfunding to continued investments by venture and angel sources.
- Increased attention to environmental, sustainability, and social issues that require new venture to support them.
- Availability of incubation facilities created under the rubric of economic development.
- A culture of startups in cluster environments like the Innovation Center in Boston. Existing real estate that has been scoped to allow and encourage this behavior.
- Enabling technology that allows faster and more cost-effective creation of prototypes and products.
- And of course, the Internet, which allows global outreach, almost instant marketing outreach, and instantaneous feedback.
- A need to create new forms of employment for early stage careers.
- The "rest of the world" is embracing innovation and entrepreneurship at breathtaking levels.

On this list of positive forces, there is the view of the entrepreneur. Some argue that the lure of high-performing and somewhat unrealistic expectations of the Facebook, Google, Amazon, and Apple are within the reach of many of the new projects. Traditionally, the public offering that results in a 10 times multiple was within expectation. It is possible that some projects will do exceptionally well, but they are the exception. Early stage investors are content when 1 in 10 investments yields higher results.

Although the expectations of upside liquidation results (the motivational "carrot") need to be in focus, the reality of early stage environment also needs clarification. There are clearly a cohort of individuals who deal with the uncertainties of the early stage companies. Those same people may not be the group to later build the enterprise. Many times inexperienced entrepreneurs express surprise at the difficulty of managing the early stage ventures because of the lack of supporting capital or people.

Some of the reality attributes include:

- Lack of recognition of either the brand or the product. This translates into longer and more expensive selling cycles. It also requires a different form of selling that relies on a term called "missionary" sales techniques where the context and need for the product (or service) needs to be identified before the sale can be considered. It's an educational process that not all sales people can bridge.
- Startup skepticism in the form of questioning the viability of the enterprise. Would General Motors (or any established organization) buy products or services from an early stage company whose very viability is questioned? Like so many specific early stage issues, this one can be countered by offering a preemptive license to the potential customer whereby they have the exclusive rights to manufacture product themselves if the startup can't. Although this option is a bit complicated, it requires a mindset that is different from an established enterprise. This, in turn, requires the early stage team to adapt in ways a more mature enterprise would not be required to embrace.
- Limited or immature organizational depth in early stage companies leads to an informal management style. Many entrepreneurs that I have spoken with have reported to be some of the best moments in their company's evolution. In my company there was a period of time when we did not have a conference table because the mill in which we were located required an odd shaped table. That forced us to have a modest but custom size table. Those of us who were there all remember the informal "standup" meetings as memorable and efficient. This was probably romanticized over time, and the frustrations due to the lack of resources was real and stood in the way of accomplishing preset goals.
- A startup was once likened to the barnstorming years of flight. Although this evokes images of the freedom of open cockpit flight, those planes cannot go very far and are considered quite unreliable. A modern Boeing 777 is capable of flying great distances and with accompanying records of significant reliability. What type of journey do you want to achieve?

A Finale

As the world's commercial, political, and economic forces change at an ever accelerating rate, so the pressures to compete in more innovative and entrepreneurial ways increase. Central governments and local municipalities have responded to these changes with a myriad of tax incentives, regulatory and legal modifications, sweeping programs such as NSF funding, and the creation of local incubation and support activities. Academia has certainly responded with its endless litany of courses, research, publications, student-run competitions, and aggressive technology transfer operations.

Enabling these efforts has been dramatic influences of enabling technology in the form of Internet-based information exchange that could only have been dreamed about a few years ago. Actual models of new ideas can be generated in real time due to computer-driven 3D modeling and the low-cost rapid prototyping. New materials allow these ideas to move to reality almost instantaneously.

Moore's Law, coined by Intel co-founder Gordon Moore in 1965, predicts that the number of transistors on an electronic chip would double each year. That trend has continued, but now the forecast is for doubling every eighteen months. That predicted dynamic has invigorated the design and implementation of electronic hardware and application software to be developed at an increasing pace anywhere in the world. Medical advances have led nearly to the ability to create genomic-based designer drugs and procedures to implement them. Robotics and automation have moved the use of these ideas to new levels of productivity.

This book looked at the process of commercial reality from the newfound wealth of new, innovative, and entrepreneurial ideas that derive from technology-based products and services. It presented a model that represents probable pathways to commercial realization of the ideas and examined the multiple forces that impact the probability of success in achieving these goals. It is certain that in order for us to successfully compete in the ever-changing world, new models and practitioners who can successfully implement them must be developed.

APPENDIX

A

Sample US Patent

US005251665A

United States Patent [19]

Schaufeld

[11] **Patent Number: 5,251,665**

[45] **Date of Patent: Oct. 12, 1993**

[54] **MECHANISM FOR OPERATING MULTIPLE AIR FLOW CONTROL VALVES**

[75] Inventor: **Jerome J. Schaufeld,** Framingham, Mass.

[73] Assignee: **Phoenix Controls Corporation,** Newton, Mass.

[21] Appl. No.: **916,906**

[22] Filed: **Jul. 20, 1992**

[51] **Int. Cl.**[5] **F16K 11/04**

[52] **U.S. Cl.** **137/554;** 137/595; 251/279

[58] **Field of Search** 137/595, 601, 607, 554; 251/231, 232, 229, 58, 279, 129.04

[56] **References Cited**

U.S. PATENT DOCUMENTS

622,114	3/1899	Burdett	137/595
677,940	7/1901	Carr	251/231 X
1,816,431	7/1931	Helf	137/601 X
3,203,446	8/1965	Smirra	137/595
3,211,177	10/1965	Phillips et al.	137/607 X
3,994,315	11/1976	Muller et al.	137/601
4,694,390	9/1987	Lee	137/554 X
4,845,416	7/1989	Scholl et al.	251/129.04 X

OTHER PUBLICATIONS

Data Sheet, Mark Hot, Inc., Arrgt Mark Air Valves Modular, no date.

Primary Examiner—Stephen M. Hepperle
Attorney, Agent, or Firm—Wolf, Greenfield & Sacks

[57] **ABSTRACT**

A mechanism is provided for operating multiple air flow control valves from a single actuator. The mechanism includes a yoke mounted to move with the drive shaft of the actuator and having a leg facing in the direction of each valve. The leg is resolvably connected, preferably by a link pivotably connected at each end, to a lever connected at its other end to a control shaft for the valve.

11 Claims, 5 Drawing Sheets

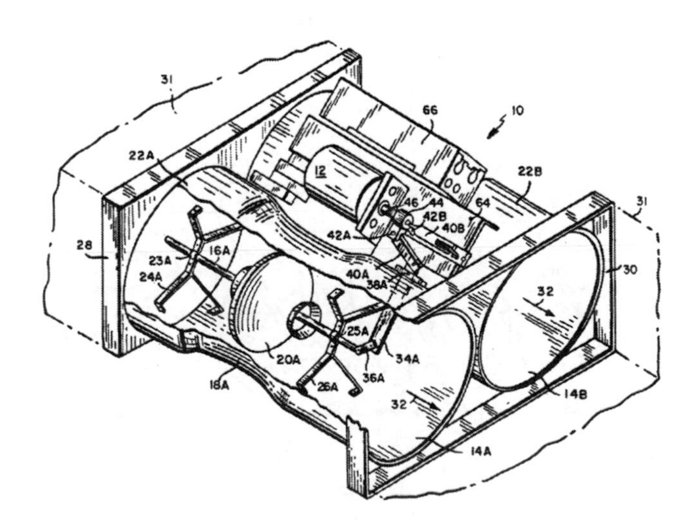

U.S. Patent Oct. 12, 1993 Sheet 1 of 5 **5,251,665**

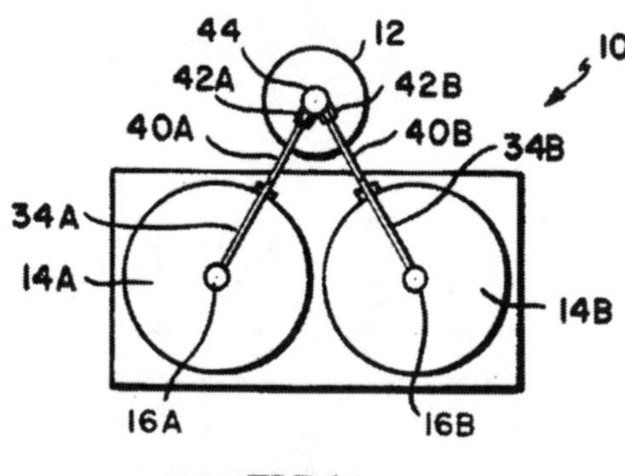

FIG.1A

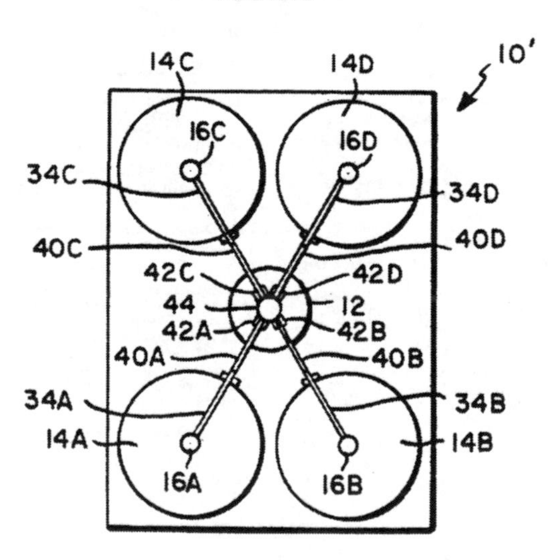

FIG.2

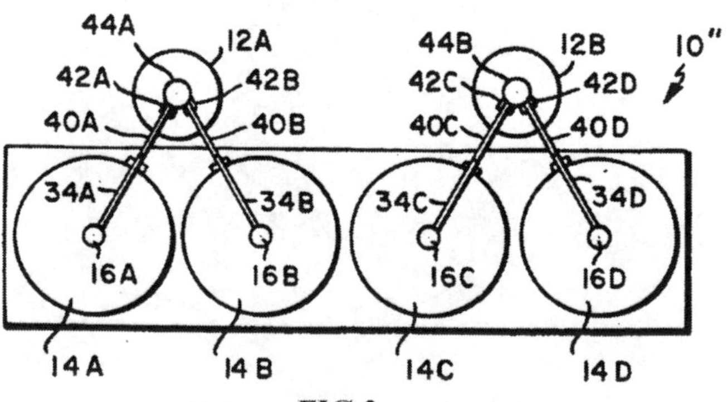

FIG.3

U.S. Patent Oct. 12, 1993 Sheet 2 of 5 **5,251,665**

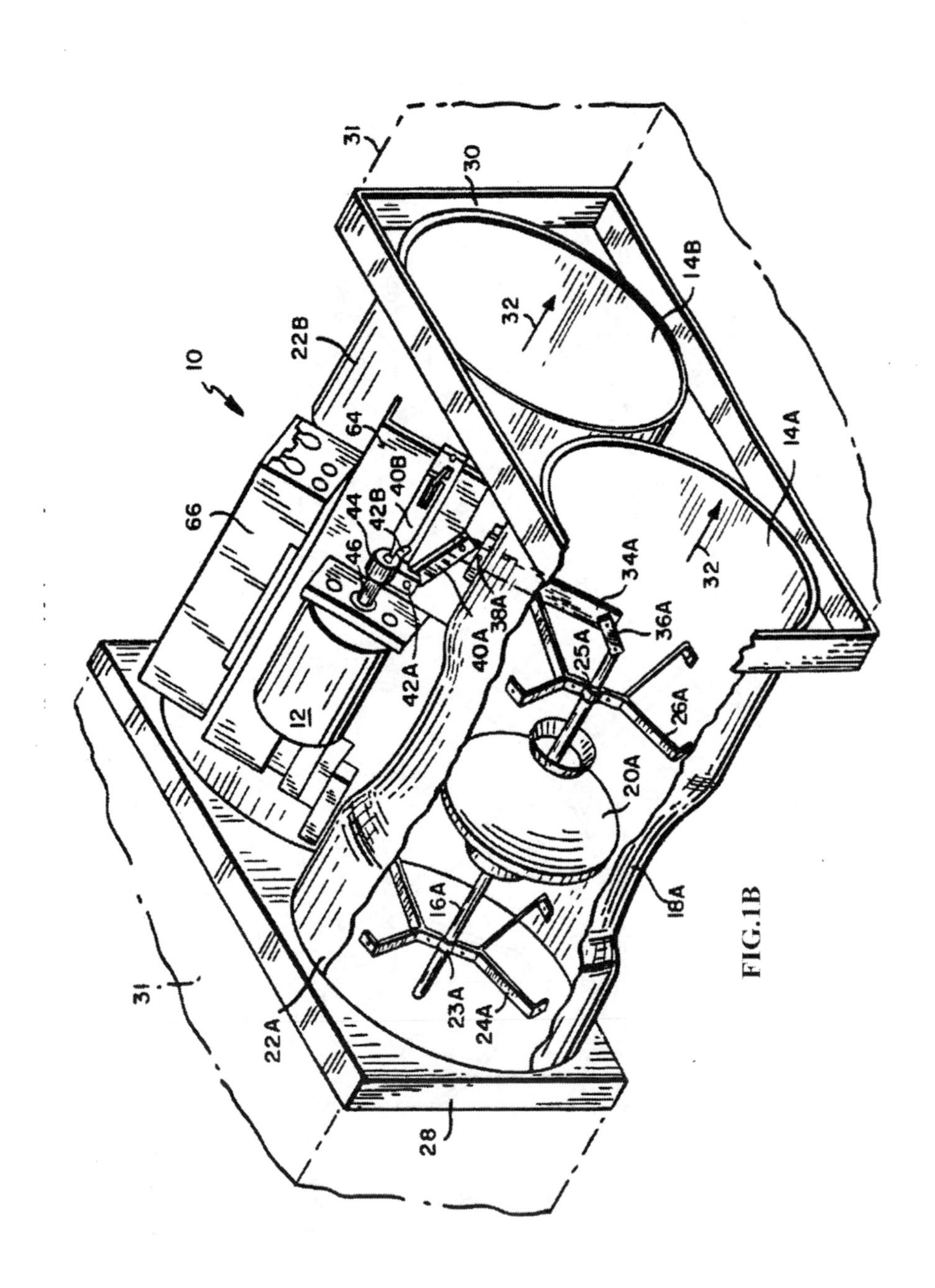

FIG.1B

U.S. Patent Oct. 12, 1993 Sheet 3 of 5 **5,251,665**

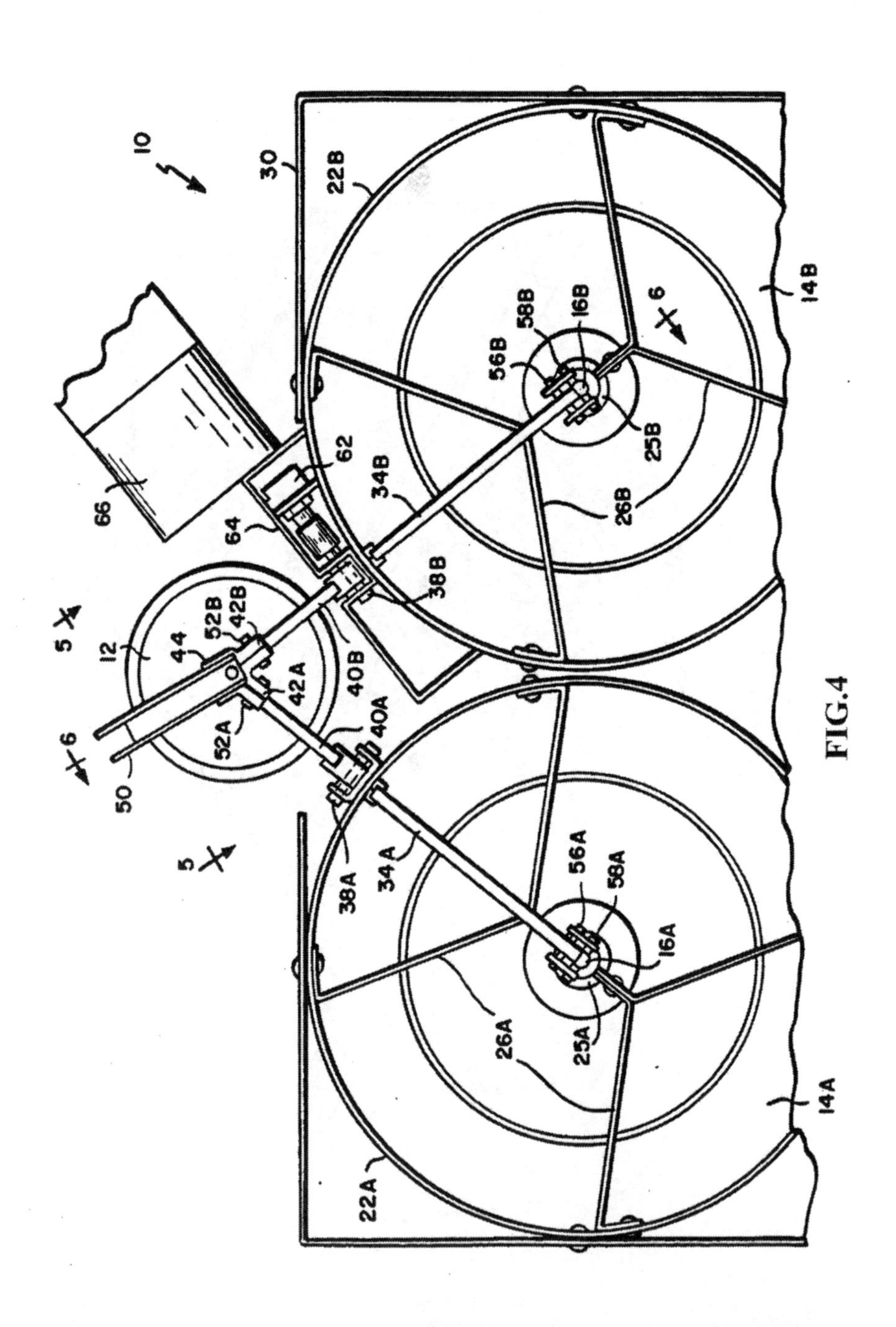

FIG.4

U.S. Patent Oct. 12, 1993 Sheet 4 of 5 **5,251,665**

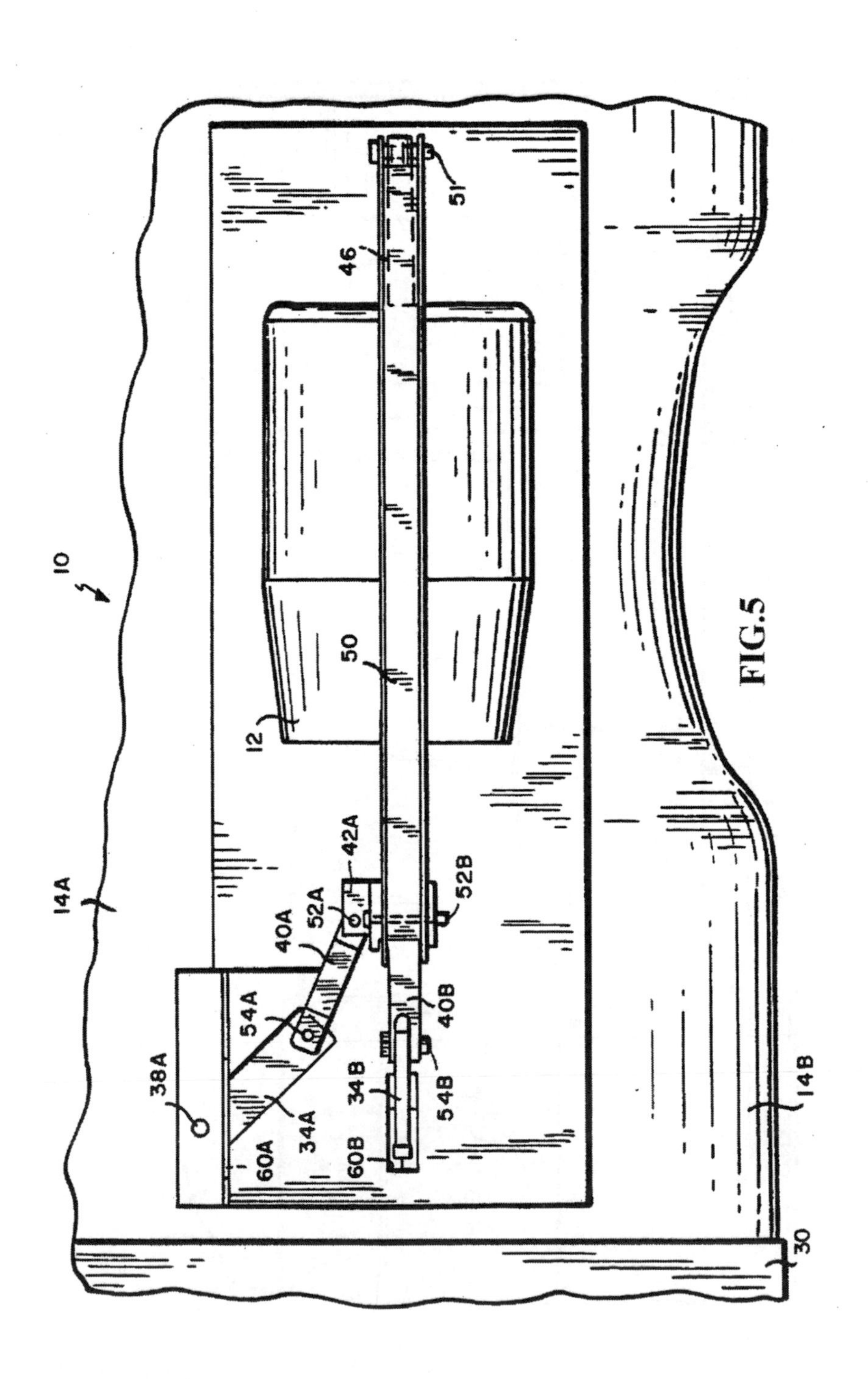

FIG.5

U.S. Patent Oct. 12, 1993 Sheet 5 of 5 **5,251,665**

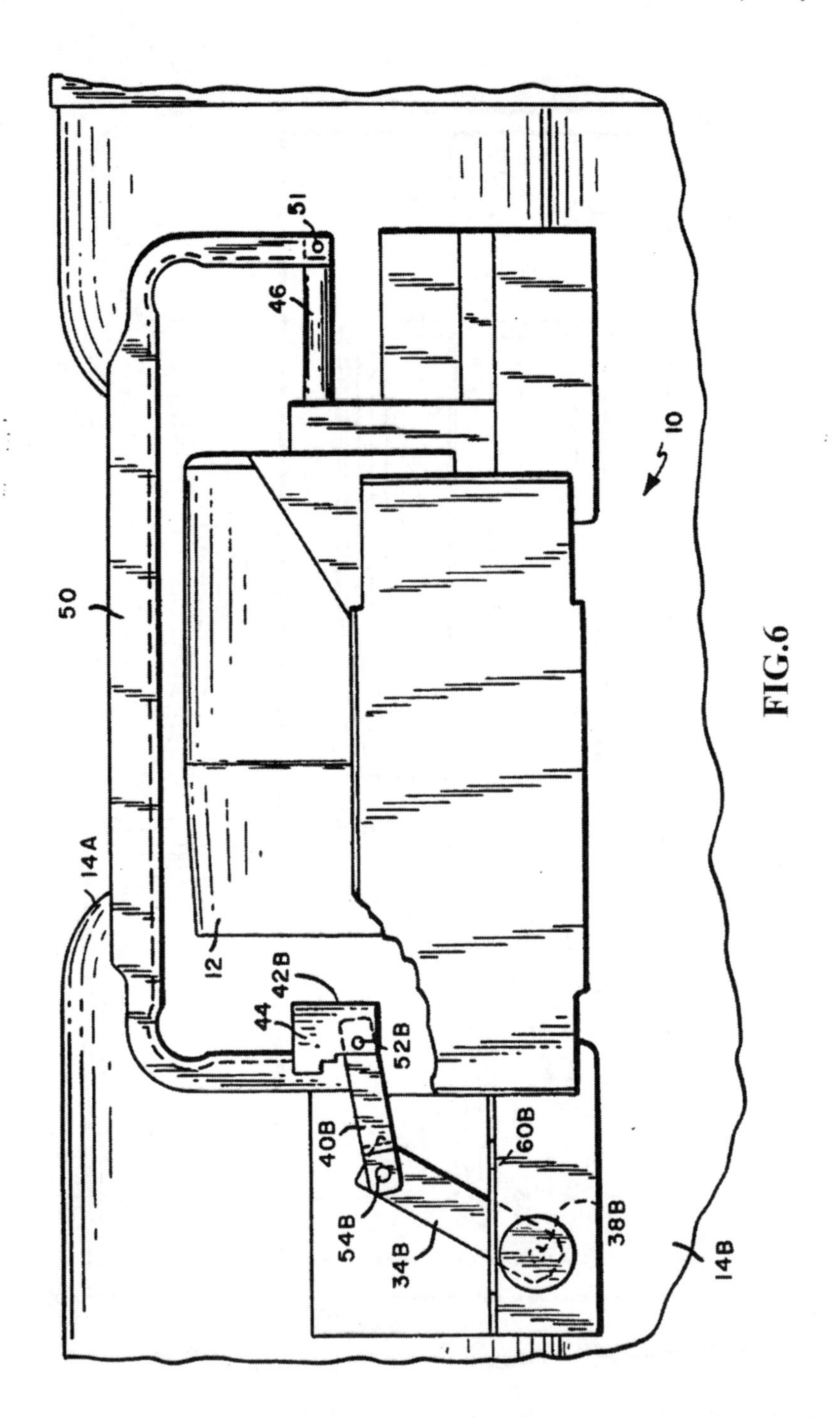

FIG.6

5,251,665

1

MECHANISM FOR OPERATING MULTIPLE AIR FLOW CONTROL VALVES

FIELD OF THE INVENTION

This invention relates to air flow control valves and more particularly to a mechanism for utilizing a single actuator to operate and control a plurality of air flow valves.

BACKGROUND OF THE INVENTION

In fume hood control systems, and in other applications where it is desired to accurately control air flow through a duct or other channel (hereinafter "duct"), Venturi or other air flow valves are frequently utilized. With such valves, there is a known and predictable relationship between the position of a valve control shaft and air flow through the valve.

While such valves provide effective air flow control, one problem has been that the ducts in which the valves are utilized are not of uniform size and shape. Since a single small valve mounted in a large duct cannot handle desired air flow through the duct, two techniques have heretofore been utilized for larger ducts, neither of which is completely satisfactory.

The first technique is to provide larger valves for the larger ducts. While this has the advantage of permitting a single actuator to control air flow in the same manner as for smaller ducts, large valves become long and unwieldly to work with. They can also be relatively heavy. In particular, because of the added length required for the larger valve, one 16 inch valve can weigh as much and take up as much area as two 12 inch valves. Since large valves also tend to magnify small errors, more precise manufacturing tolerances are required for such valves, making them more difficult and expensive to manufacture. In addition, Venturi valves normally have a generally circular cross section and therefore generally fit well in ducts which are substantially square or circular. However, because of space limitations in buildings, large ducts are typically rectangular, making it difficult to fit a large valve in such ducts. Finally, it is substantially less expensive to build one or two valve modules in selected sizes and to gang such modules to fill larger ducts than it is to build valves in all of the various sizes which might be required for different duct sizes.

However, ganging two or more smaller valves to fill a duct opening also presents some problems. One problem is that, with current designs, the position of each valve needs to be independently measured and a separate actuator provided to drive the valve. Separate electronic feedback controls are also required for each valve/actuator pair. This arrangement, with multiple actuators, is expensive, and is difficult to install and maintain. It also requires additional testing and calibration to assure that the various valve/actuator combinations are operating in synchronism.

This suggests that the advantages of a multi valve system for large ducts can be achieved without the disadvantages by providing a single actuator and a single control for all of the valves (or at least for selected numbers of the valves). In doing this, advantage is taken of the fact that the air flow through a given valve for a given valve shaft position is a constant which can be determined, so that if the shaft position for one valve is measured and the valves can be tied together so as to operate in tandem, it should be possible to use the single

2

valve measurement to control the actuator driving all of the valves. However, in practice, it has been found difficult to design a mechanism for linking the valves to the actuator in a manner such that the valves move in tandem. In particular, the moment arms of such linkages tend to permit linkage skewing, resulting in uneven movement of the valves and thus in poor control of air flow. While efforts have been made to stiffen the linkages between the actuator and the valves, such efforts have not heretofore been completely successful, the linkages either not being stiff enough or being too bulky and cumbersome to be used effectively. As a result, use of ganged valves from a single actuator has not heretofore been commercially practical.

A need therefore exists for an improved mechanism for linking a single actuator to operate a plurality of valves, which mechanism assures that the valves move in tandem so that a valve position measured for one of the valves can be utilized to control the actuator to provide a desired precise air flow through the ganged valve combination.

SUMMARY OF THE INVENTION

In accordance with the above, this invention provides a mechanism for converting substantially linear movement of a valve actuator into predictable proportional, substantially linear movements of at least two air valve control shafts. The mechanism utilizes a yoke attached to move with the valve actuator, which yoke has a leg for each valve, each leg extending generally in the direction of the control shaft for the corresponding valve. There is also a lever for each valve which is pivotally connected at one end to the corresponding valve control shaft and extends in at least on plane toward the corresponding yoke leg. The yoke leg and the lever for a given valve are resolvably connected. For a preferred embodiment, this connection is a link pivotably connected at one end to the yoke leg for the corresponding valve and pivotably connected at the other end to the extending end of the corresponding lever.

Each lever is preferably pivoted at a selected point between its ends so that the valve shafts are moved by the actuator in a direction opposite from the direction of actuator movement The relative lengths of the yoke legs, the levers, and the links for each valve should be such that there is a predetermined ratio between the movement imparted to each valve shaft as a result of a given actuator movement. For the preferred embodiment, the movements of the valve shafts for a given actuator movement are identical.

Preferably, the yoke leg, link and lever for each valve are, when viewed in the direction of the valve shaft, in substantially the same plane. The axis of the actuator and of the valve shafts are all substantially parallel for preferred embodiments, and either two valves or four valves are controlled from a single actuator. Shaft position is measured for one of the valve shafts and means are provided for utilizing the measured position to control the actuator for operating all of the drive shafts. For preferred embodiments, measurement of shaft position is accomplished by measuring the position of the corresponding lever and the valve is a large air volume, low pressure valve such as a Venturi valve.

The foregoing and other objects, features and advantageous of the the invention will be apparent from the following more particular description of preferred embodiments of the invention as illustrated in the accompanying drawings.

IN THE DRAWINGS

FIG. 1A is a diagrammatic front sectional view showing an actuator being utilized to control two valves in a rectangular duct.

FIG. 1B is a partially cut away front top perspective view of a one actuator two valve assembly of the type shown in FIG. 1A.

FIG. 2 is a diagrammatic front sectional view showing a single actuator being utilized to control four valves in a large, substantially square duct.

FIG. 3 is a diagrammatic front sectional view of a pair of actuators being utilized to control four valves in a large rectangular duct.

FIG. 4 is an enlarged front view of a two valve, one actuator preferred embodiment.

FIG. 5 is a top view generally looking in the direction of arrow 5 in FIG. 4.

FIG. 6 is a side view generally looking in the direction of arrow 6 in FIG. 4.

DETAILED DESCRIPTION

Referring to FIGS. 1A and 1B, a valve assembly 10 is shown having a single actuator 12 and two valves 14A and 14B. Each valve 14 will be assumed to be a standard Venturi valve of a type currently available from Phoenix Controls Corporation, 55 Chapel Street, Newton, Massachusetts. Such valves are preferably low pressure (for example 0.6" to 3" water), large air volume (for example 60 cfm to 5,000 cfm) valves. However, the invention is not limited to use with such valves, and may be employed in other applications where air flow through a valve varies as a function of the position of a valve control shaft (16A,16B).

As may be best seen in FIG. 1B, each Venturi valve 14 has a reduced diameter portion (18A,18B) near its center and has a generally cone shaped plug (20A,20B) mounted to move with shaft 16. As plug 20 moves into reduced area 18, it reduces the size of the valve orifice, and thus reduces air flow through the valve. As plug 20 is moved out of reduced area 18, the air flow area of the valve increases, permitting greater air flow through the valve. Each shaft 16 is supported in the corresponding valve housing 22 by bearings (23A,23B) and (25A,25B) supported by brackets (24A,24B) and (26A,26B), respectively, the shaft being slidable forward and backward in the bearings.

Valve assembly 10 has a pair of end plates 28 and 30, with flanges on the plates which are adapted to mate with edges of the duct 31 in which the valve assembly is mounted to seal the duct so that air can only pass through valves 14. As viewed in FIG. 1B, the direction of air flow is indicated by arrow 32 pointing from plate 28 to plate 30.

The position of shaft 16, and thus of plug 20, in each valve 14 is controlled by a lever arm (34A,34B) which is pivotably connected to the corresponding shaft 16 through a linkage (36A,36B) and is pivoted at a pivot (38A,38B) which is on the outer edge of the corresponding housing 22. This linkage will be discussed in greater detail in conjunction with the description of FIGS. 4-6. As will also be discussed in greater detail later, the upper end of each lever 34 is connected through a link (40A,40B) to a corresponding leg (42A,42B) of a yoke 44. Yoke 44 is attached to move with the drive shaft 46 of actuator 12. In FIG. 1B, this connection is a direct connection, while in other figures the connection is indirect. As may be seen in FIG. 1A, and is better seen in FIG. 4, the element 34, 40 and 42 for a given valve are in substantially the same plane when viewed along the direction of air flow.

The various parts of the valve housing, plates, and other components may be of material commonly used for such parts including steel, aluminum, and the like.

FIG. 2 shows an alternative embodiment 10' of the invention wherein yoke 44 has four legs 42A-42D, each of which leads to corresponding links 40 and levers 34 for a corresponding valve 14. Except for the difference of using four valves connected to operate off a single actuator rather than two valves, the embodiment of FIG. 2 is substantially the same as that shown in FIGS. 1A and 1B and operates in substantially the same manner. FIG. 3 shows a second alternative embodiment 10' for the valve assembly wherein two valve assemblies of the type shown in FIG. 1A are mounted side by side in a single valve assembly to provide a two actuator, four-valve embodiment. While this embodiment still provides some of the calibration problems associated with having multiple actuators, it presents less problems than for a configuration of this type employing four actuators.

FIGS. 4, 5 and 6 are more detailed views for the embodiment of the invention shown in FIGS. 1A and 1B. In particular, FIG. 4 is an enlarged and more detailed version of FIG. 1A. Like elements in all of the figures have been given the same reference numeral. The only difference between the embodiment shown in FIGS. 1A and 1B and that shown in FIGS. 4-6 is that yoke 44 in FIG. 1B is attached directly to shaft 46 so that when actuator 12 moves outward, the yoke, and thus the upper portion of lever arms 34, are moved to the right as shown in FIG. 1B or out of the paper as shown in FIG. 1A (or FIG. 4). This results in the corresponding cone or plug 20 being moved to the left to open the corresponding valve. Conversely, in FIGS. 4-6, the yoke 44 is attached at the end of one arm of a U-shaped bracket 50, with the end of the other leg of the bracket being attached to shaft 46. Thus, when shaft 46 is extended by the actuator, yoke 44 and the upper ends of lever arms 34 are driven into the paper as shown in FIG. 4 or to the right as shown in FIGS. 5 and 6, resulting in cones 20 being moved into corresponding throats 18 to reduce air flow through valves 14. Thus, while in each instance the valve shafts and the cones or plugs affixed thereto move in the opposite direction from the direction in which the actuator shaft is being moved, the direction of valve shaft movement, and thus the opening or closing of the valve is reversed for an actuation of the actuator.

In addition, FIGS. 4-6 more clearly show that each link 40 is connected to the corresponding yoke arm 42 by a cotter pin (52A,52B) and that each link 40 is connected to the corresponding lever arm 34 by a corresponding cotter pin (54A,54B), respectively. Similarly, there is a cotter pin (56A,56B) linking the lower end each lever 34 to the corresponding link 36 and a cotter pin (58A,58B) linking the other end of each link 36 to the corresponding valve shaft 16. The pivot 38 for each lever 34 is shown in more detail in FIGS. 4-6 as being a bolt passing through the lever arm, which bolt is mounted to the top of housing 22. A slot (60A,60B) is provided in the top of each housing 26 through which the corresponding lever arm 34 passes, the slot being long enough to permit movement of the lever arm from, for example, the position shown in FIG. **5** to a position approximately 90° to the left of such position.

The rotary position of lever **34B** is detected by a detector **62** mounted to a bracket **64** and attached to the lever junction **38B**. Signals from detector **62** are applied to an electronic flow controller circuit **66** which, in response to the outputs from detector **62**, generates signals to actuator **12** to control the position of shaft **46** and thus the position of cones **20** in valves **14**. Thus, a single position detector is utilized to control all flow valves being operated from a single actuator **12**. This can be done since there are no linkages having moment arms between valve shafts, the drives for all valve shafts being linked at a common point to the actuator shaft. Thus, all valve shafts move in tandem so that a measurement of the position of one valve shaft is indicative of the position of all valve shafts.

While for the preferred embodiments, the movement of all valve shafts **16** is the same for a given movement of actuator **12**, this is not a limitation on the invention. Thus, by adjusting the relative lengths of yoke arms **42**, links **40** and levers **34**, two valves **14** of different size might be utilized, with the relative movement of the two valve shafts being constant, for example, 3 to 2, but not being equal.

Further, while the actuator shaft **46** and the two valve shafts **16** have all been shown as being parallel for the preferred embodiment, this is also not a limitation on the invention. Therefore, the invention could be practiced, with, for example, the end of the actuator shaft not in contact with yoke **44** being higher or lower than the yoke end of the shaft, or the end of bracket **50** in contact with yoke **44**, while still remaining within the teachings of the invention. It is also possible that in certain applications the end of one or both yoke arms **42** could be connected directly to the end of lever arm **34**, rather than through intermediate link **40**. However, since the relative movement at this junction involves movement in at least two dimensions (i.e., up and down as well as forward and backward), such linkage would have to be in a cam track or other resolvable linking mechanism which would provide the freedom of action for such relative movements. It is also possible to alter the nature of the various linkages and of the specific components used or of their inter relationship for specific applications. The ducts **31** in which the valve assemblies are mounted have also been assumed to be rectangular for the preferred embodiments. However, this invention has the flexibility to be used with ducts which are oval, circular, or having other shapes by use of a suitable number of valves of appropriate size and appropriately shaped end plates **28** and **30**.

Thus, while various embodiments of the invention have been described in some detail above and various possible modifications on such embodiments have been mentioned, it will be apparent to those skilled in the art that these and other changes in form and detail ma be made by those skilled in the art without departing from the spirit and scope of the invention.

What is claimed is:

1. A mechanism for converting substantially linear movement of a valve actuator into predictable, proportioned, substantially linear movements of at least two air valve control shafts, the mechanism comprising:

a yoke attached to move with said valve actuator, said yoke having a leg for each valve, which leg extends generally in the direction of the corresponding valve control shaft;

a lever pivotably connected at one end to the corresponding valve control shaft and extending in at least one plane toward the corresponding yoke leg; and

means for resolvably connecting the yoke leg for each valve to the extending end of the corresponding lever, each means for resolvably connecting including a link pivotably connected at one end to the yoke leg for the corresponding valve and at the other end to the extending end of the corresponding lever.

2. A mechanism as claimed in claim **1** wherein said lever is pivoted at a selected point between its ends, whereby the valve shafts are moved by the actuator in a direction opposite from the direction of actuator movement.

3. A mechanism as claimed in claim **2** wherein the relative lengths of the yoke legs, the levers and the links for each valve are such that there is a predetermined ratio between the movements imparted to each valve shaft as a result of a given actuator movement.

4. A mechanism as claimed in claim **3** wherein the movements of the valve shafts for a given actuator movement are identical.

5. A mechanism as claimed in claim **1** wherein, when viewed in the direction of the valve shaft, the yoke leg, link and lever for each valve are in substantially the same plane.

6. A mechanism as claimed in claim **1** wherein the axis of the actuator and of the valve shafts are substantially parallel.

7. A mechanism as claimed in claim **1** wherein two valve shafts are controlled from each actuator.

8. A mechanism as claimed in claim **1** wherein four valve shafts are controlled from each actuator.

9. A mechanism as claimed in claim **1** including means for detecting the position of one of said valve shafts, and means for utilizing the detected position to control said actuator.

10. A mechanism as claimed in claim **9** wherein said means for detecting measures shaft position by detecting the position of the corresponding lever.

11. A mechanism as claimed in claim **1** wherein said air valve is a Venturi valve.

* * * * *

I

Index

Apress®
THE EXPERT'S VOICE™

CPSIA information can be obtained
at www.ICGtesting.com
Printed in the USA
LVOW10s0025020218
564987LV00008B/68/P